I'm a European

7 Look at the map of Europe on the left and at the sentences below. What can you say about European countries and cities? Write the sentences in your exercise book and add lots of details.

- I'm from …
- My relatives live in …
- I've got friends in …
- I've (often) been to … (once/twice)
- I know a lot / a bit about … (because …)
- I'd like to visit … because …

8 Now copy the map on the left and mark all the places that you've just talked about on the copy.

A quiz on Europe

9 Find the "Odd word out". Explain your answer.

1	Netherlands	Canada	Denmark	Portugal
2	Birmingham	Paris	Berlin	Madrid
3	Rhine	Thames	Mississippi	Danube
4	Slovakia	Majorca	Cyprus	Sicily
5	Scotland	England	Luxembourg	Wales
6	Stockholm	Brussels	Helsinki	Oslo
7	Germany	Austria	Switzerland	Poland

1. Canada is wrong here because it isn't a European country.
2. …

10 Make your own quiz on Europe with a partner. Find four questions. Then swap quizzes with other groups in your class.

A project on Europe

11 Start a project on a European country with two or three classmates.

a Draw a map of the country. Mark the capital and some other cities on the map. Mark some rivers and mountains, too.

b Give some information about two or three interesting places in the country and one or two important events from the past.

c Write down what you think is typical for the country.

d Add the country's flag and lots of photos.

You can present your ideas as a poster or as part of your school's website, for example. Explain them to your class.

Welcome to Berlin

12 **What do you know about Berlin? Collect all your ideas about Berlin in a network. You can start like this:**

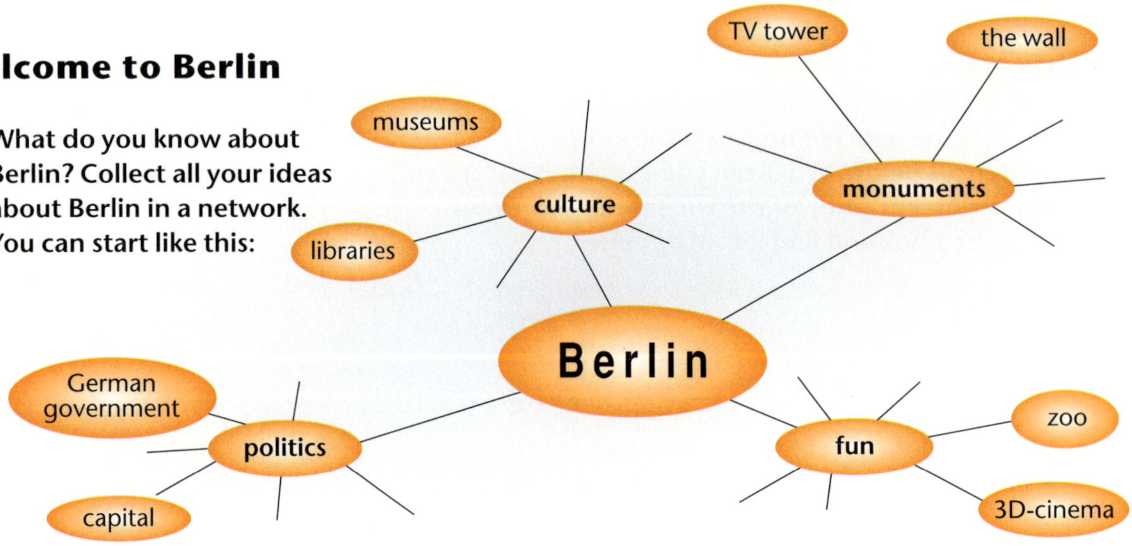

13 **Look at these photos (1–7) from Berlin and match them to the sentences (a–g) below.**

a It opened on 1st August, 1844. It has 19,000 animals.
b It was on the border between East and West Berlin until November 1989. It's now a symbol of the united Germany.
c It's the biggest department store on the European continent. It has 80,000 visitors per day.
d It's probably from 1340 BC. It's made of chalk and is only 50 cm high.
e It's 368.03 m high. It's 986 steps if you walk. There's a café at 207.53 m.
f It opened in 1998. Its screen is 20 by 27 m and it has 24,000 watt sound.
g It's the home of the German government. It's famous for its new glass dome.

Helpful expressions

- You write: 50 cm
  You say: fifty centimetres
- You write: 368.03 m
  You say: three hundred and sixty-eight point o three metres

▸ p. 55

14 **What are the names of the places and things in task 13?**
A library, a travel agency or the Internet can help you.
Helpful website: http://www.berlin.de/home/English

15 **Which of the places in task 13 would you like to visit? What other place(s) in Berlin would you like to see? Explain your answers.**

Welcome to Warsaw

Dear Alexander,

My name is Jan. I'm 13 years old and I'm looking for a pen-friend in Germany. I can't speak German, so I'm writing this letter in English. I hope you can speak English, too. My favourite subject is Geography, so in this letter I'm going to tell you something about my hometown, Warsaw.

Warsaw ("Warszawa" in Polish) is Poland's biggest city. More than 1.6 million people live here. Most Polish people are Catholics. A very famous Polish person is Pope Paul II, but he lives in Rome, of course. Most buildings in Warsaw today are no more than 50 years old, because about 85% of the city was destroyed at the end of World War II. A lot of people had been deported or killed. Before the war more than 30% of Warsaw's people (then also about 1.6 million) were Jewish.

Warsaw has a lot of wide streets, but visitors often say that the city is very green. They also like the historic buildings. The buildings along the "Royal Route" have been reconstructed, and the Old Town is now a UNESCO World Heritage site.

Most visitors come to our city from May to October, when the weather is nice. The temperature often rises above 30°C. But January and February are very cold months and the temperature can fall to -30°C. Do you have cold winters, too?

I'm sending you some photos of Warsaw. I hope you like them. Write back to me and tell me something about where you live. Maybe you can send me some photos, too.

Hope to hear from you soon!

Yours,

Jan

Frederic Chopin, the great pianist, came from Warsaw.

Warsaw street

Royal Castle

Warsaw's Old Town

16 Read Jan's letter and write about …

a how big Warsaw is.
b most Polish people's religion.
c Warsaw's history after World War II.
d Warsaw's Old Town.
e the weather in Warsaw.

17 Imagine you want to write a letter back to Jan. Tell him about your hometown.

Helpful expressions

- You write: 1.6 million
  You say: one point six million
- You write: 30°C or -30°C
  You say: (minus) thirty degrees Celsius

Chapter 2

Pompeii – a town under a volcano

1 Look at the picture on the left – what do you think is happening here?

2 Draw lines on a copy of the picture so that you've got nine parts.
Then write down these words:

Now say what you can see in one of the parts – the others must say what part that is.

I can see two windows in my part.

Is that at the top on the left?

▸ p. 41

3 Look at the girl in the foreground on the right. Imagine it's you. Copy this star into your exercise book and write down around it what she sees, feels, smells, etc.

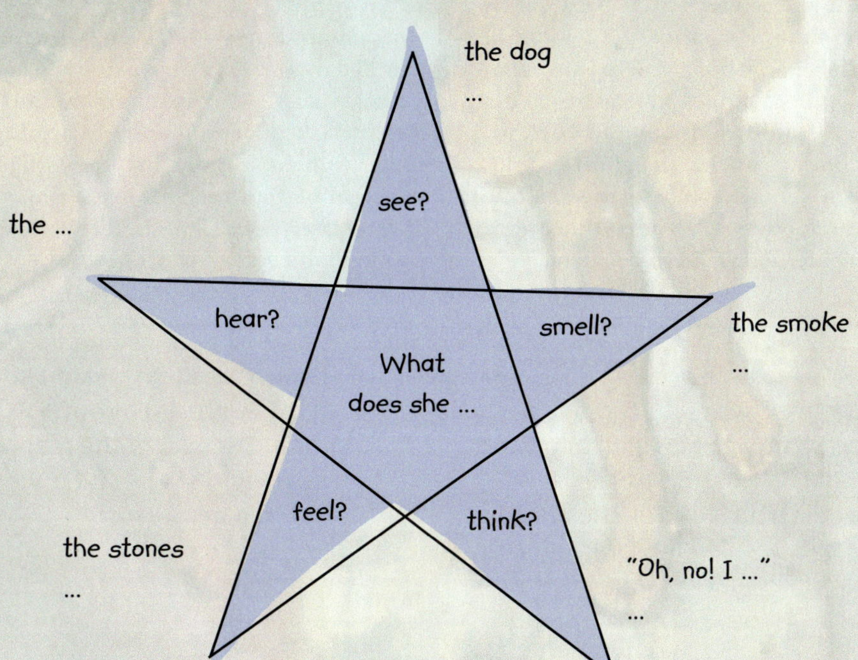

4 Look at the map and copy it into your exercise book. Then:

a Colour the land brown and the sea blue.

b Mark and label the following places: Mount Vesuvius, Misenum, Herculaneum, Stabiae and Pompeii.

c How far are the four towns from Mount Vesuvius?

d Look at the key: What do the three colours mean?

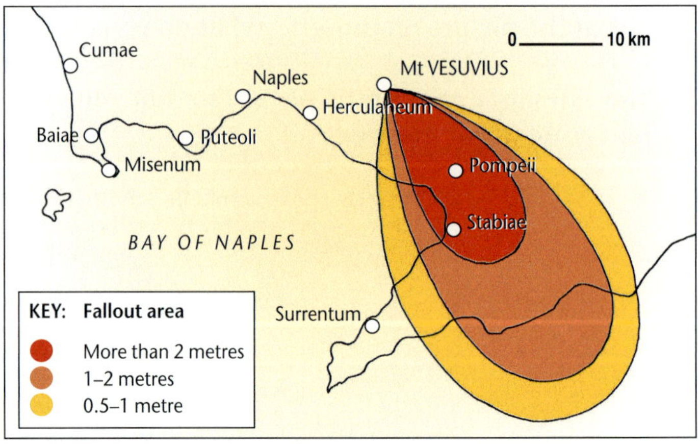

5 Read this true story of the day when Vesuvius erupted. Publius wrote it down later.

In the year 79 the admiral of the Roman fleet, Caius Plinius Secundus, was spending the summer in Misenum with his sister and her 17-year-old son, Publius Caecilius Secundus. Everything was great, but on 24th August …

Helpful vocabulary

branch [brɑːntʃ] *Ast* • fear [fɪə] *Angst*
shore [ʃɔː] *Ufer* • pillow ['pɪləʊ] *Kissen*
weak [wiːk] *schwach*
breathing ['briːðɪŋ] *Atmen*

The vocabulary list (p. 50–55) has all the other new words. Tip: Try to guess the words first.

Early in the afternoon, my mother showed my uncle a strange cloud. It looked like a very tall tree with branches only at the top. In some places the cloud was white, in others a dirty colour.

My uncle wanted to see it better and got a boat ready. But just when he was leaving, he got a call for help from people near Herculaneum. So he put the warships to sea to help them.

Everyone was leaving the danger zone; my uncle showed no fear and sailed into it. All the time, he watched each new phase of the eruption; a slave wrote down all the details for him. Now ashes were falling around them, more and hotter when they came nearer the shore, together with stones. Suddenly they could not go on because the shore was full of stones and ashes. The people on the ship wanted to go back to Misenum, but my uncle, the admiral, told them to turn south to Stabiae, to save his friend Pomponianus. He landed there and took his frightened friend in his arms.

Then, to show other people that they should have no fear, he went to have a bath and dinner. All the time on Vesuvius, the fires made the sky bright and great flames jumped out on all sides. But my uncle went to bed.

When there were more and more ashes and stones, the people woke my uncle up to save him and discuss what to do. The buildings were shaking, so it wasn't safe to stay in the house, and outside, stones were falling from the sky.

In the end, they put pillows on their heads and went down to the shore. They wanted to sail away by boat, but the sea was wild and dangerous.

It was daytime now, but near Vesuvius it was still dark – darker than a normal night. My uncle was weak from the smoke and the ashes. He asked for cold water to drink. His friends wanted to help but he collapsed and the smoke stopped his breathing.

36 hours later, my uncle's body was found. He didn't look dead, but he was.

6 Find out …

a why Publius' uncle wanted to go out in a boat.

b why he took some warships with him later.

c what Publius' uncle did in Stabiae.

d why the people put pillows on their heads.

e why they couldn't sail away from Stabiae.

f what killed Publius' uncle.

7 Look at the text again and draw Plinius' sea route on the map. Then label the map with notes and say what happened at the places.

Volcanoes

8 What happens when a volcano erupts?

a Copy this cross section of an erupting volcano. Colour it and label the different parts: magma, vent, crater, lava, volcanic ash, … You can go to a library or use the Internet to get help. Helpful website: http://volcano.und.nodak.edu/vw.html

b Now try to say what happens when a volcano erupts.

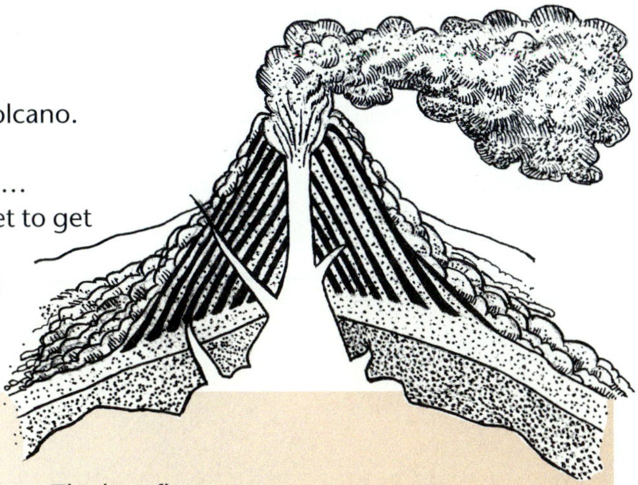

Helpful expressions

- The … builds up pressure.
- … erupts and shoots … into the sky.
- lava, gases, stones and ashes
- The lava flows …
- Fertile ground around the volcano …

9 Find out about a volcano (not Vesuvius).

a Where is it?

b Is it still active?

c When did it last erupt?

d What was the effect of the eruption on the countryside?

10 Back to Vesuvius. Try to find out more about the eruption of Vesuvius in the year 79.

a How long did the eruption last?

b What was the result for the area around the volcano?

c Has Vesuvius erupted again since the year 79?

A Roman town: Pompeii

After the eruption 2,000 of Pompeii's 20,000 people were dead and the town was buried under more than two metres of ashes. It was still buried 1,700 years later. But then people began to dig out the remains of the town. And lots of it looked just like on 24th August in the year 79. Today thousands of visitors go through the streets that were covered in ashes so long.

The photo on page 11 and the map below show the same things.

11 Find these places on the map:
Town wall with guard towers • Forum • Hall for the cloth trade • Little market hall
Temple of Jupiter • Temple of Apollo • Theatre • Barracks for gladiators • Amphitheatre
Palaestra • Central baths • Cemeteries

Mark them on a copy of the map.

1 Stone wall with square guard towers and gates, surrounding the town
2 Forum – the town's main square, surrounded by public buildings
3 Town's offices
4 Hall for the cloth trade
5 Basilica
6 Comitium
7 Macellum or little market hall
8 Temple of Jupiter
9 Temple of Apollo
10 Temple of Vespasian
11 Temple of the Lares
12 Triangular forum – the earlier town centre
13 Temple of the Egyptian goddess Isis

PUBLIC ENTERTAINMENT

14 Theatre
15 Odeon or little theatre
16 Barracks for gladiators
17 Amphitheatre
18 Palaestra – open courtyard for exercise, surrounding a swimming-pool

PUBLIC BATHS

19 Baths near the forum
20 Central baths
21 Baths on the street to Stabiae
22 Water reservoir

HOUSES

23 House of the Tragic Poet
24 House of the Vetii Brothers

25 Cemeteries outside the town wall

12 Now try to find the same things on the photo on page 11. Mark them on a copy. What differences can you see between the map and the photo?

13 Look again at the town and its buildings. What things were important for the Romans? Write them down.

Blow your top

14 What's it like to be near an erupting volcano?
This is what you can do to make your own volcano.

You need:

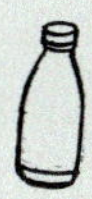
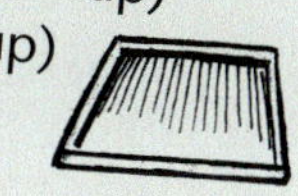

- sand or dirt
- water (1/2 cup)
- a metal can (small, thin)
- a small bottle
- baking powder (4 tablespoons)
- red food colouring (if you like)
- washing-up liquid (1/4 cup)
- white vinegar (1/4 cup)
- a big baking sheet

Then:

- Put your baking sheet on a table.
- On the sheet, make a mountain of sand around the metal can, but leave the top of the can open. This is the crater of your volcano. (It's easier to make the mountain when the sand is a little wet.)
- Pour the baking powder into the metal can.
- In the small bottle, mix the water, washing-up liquid and vinegar. (Add some red food colouring if you like.)
- Pour only a little of the mixture from the bottle into the mountain's "crater hole". (You can use the rest of the mixture later to make your volcano erupt again.) Step back and watch your volcano erupt! The "lava" will flow over the crater and down the sides of the mountain, just like it would from a real volcano.

Tip:
If your volcano doesn't erupt right away, stir the mixture in the can with a metal stick.

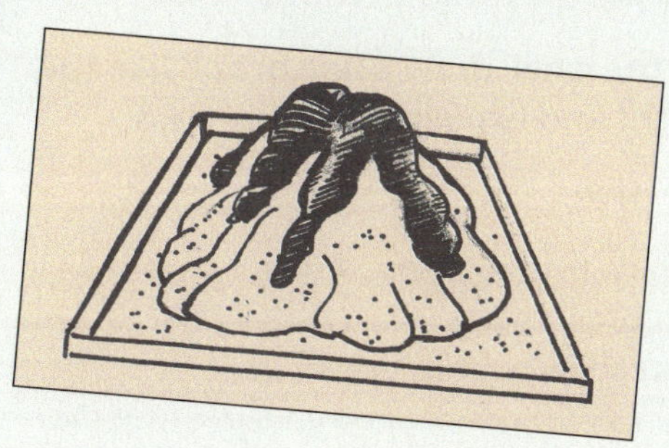

The weather

Think about the weather around you. What's it like at the moment? What was it like earlier? What do you think it might be like later?

1 Look at the photos below. Write down what effect the weather has on the people in the photos.

2 Explain the effects that the weather can have on people with different jobs. Think of:

- strong headwinds – a pilot
- heavy rain – a taxi driver
- frost – a builder
- snow – the manager of a hotel in the mountains
- heat and sun – the manager of a supermarket
- ice and snow – a car driver before a trip

Now draw pictures to show the effects. Think of a good caption for each drawing.

3 For about a week, collect news reports that have something to do with the weather. Before you put them on a wall in your classroom, explain them to your classmates.

Helpful expressions

- This report is about … / This is a report on …
- I've found a report that says that there was … in … yesterday/…
- heavy rainfall/snowfall / a thunderstorm / a hurricane / a heat wave / a drought / a flood / …
- There was a road accident after heavy fog / a traffic jam after heavy snowfall / a flooded area / …

4 Collect words that describe the weather and make a network. Think of:
cold, sunny, … • heavy rain, bright sunshine, … • windy, moderate, … • cloudy, clear, …

Make a big collage in your classroom with photos or drawings for your weather words.

5 Find out about the following weather elements and explain them:
temperature, precipitation, wind speed, wind direction, air pressure, clouds.
Helpful website: http://www.wxdude.com/basics.html

6 For a week, use a thermometer to measure and write down the air temperature every day like this: Monday: 9°C; Tuesday: …

At the end of the week draw a line graph like the one on the right and explain it to your class.

Helpful expressions

- (-)5°C = (minus) five degrees Celsius
- The temperature remains level / rises / falls.
- It rises/falls steadily/quickly/slowly/…

▶ p. 42

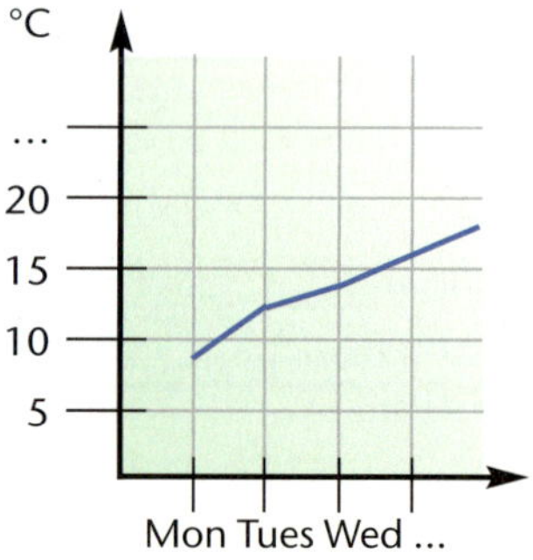

Activity

7 **Look at the barometer on the right. Make your own barometer and find out if the air pressure is rising (high) or falling (low).**

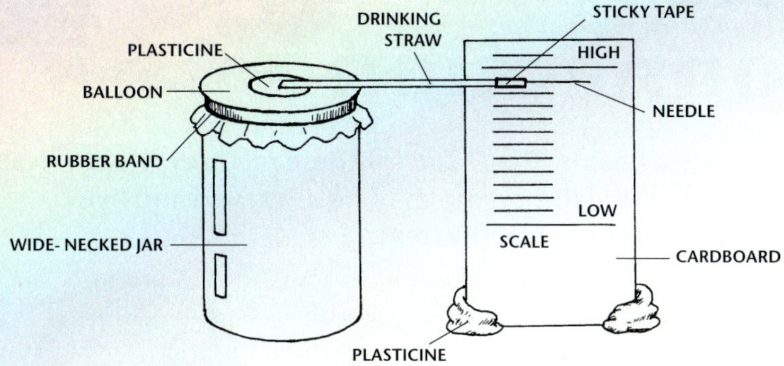

Weather records

- Highest world temperature: 58°C, Al Aziziyah, Libya, 13th September, 1922
- Lowest world temperature: -68°C, Oymyakon, Siberia, 6th February, 1933
- Greatest rainfall in a day: 73.62 inches, Réunion, Indian Ocean, 15th March, 1952
- Greatest rainfall in a year: 1,041 inches, Assam, India, August 1880–1881
- Fastest tornado winds: 286 miles per hour, Wichita Falls, Texas, 2nd April, 1958

8 **Read the above weather records correctly. Then find the places in an atlas and find out why it's**
a) so hot in Libya; b) so cold in Siberia;
c) so rainy in Réunion and in Assam in India.

Now find places in Germany that are among
a) the coldest; b) the sunniest; c) the rainiest.

Helpful expressions

- 73.62 inches = seventy-three point six two inches
- 1,041 inches = one thousand and forty-one inches

(1 inch = 2.54 cm; 1 mile = 1,609 km)

▸ p. 55

The water cycle

Water is an important part of our ecosystem. Living things need it to survive. Water evaporates from the oceans into the atmosphere. The atmosphere then carries the vapour elsewhere, until clouds and precipitation are formed and the rain or snow falls to the ground. Much of this water then flows back to the oceans through rivers and starts the cycle all over again.

9 **Look at a map of Germany in an atlas. Find three rivers that carry water to the oceans. Then find mountains or hilly regions in Germany, where a lot of water falls as rain or snow.**

10 **Look at the diagram above. Match the numbers in the diagram (1–4) to these sentences:**
a The water evaporates and rises into the air as water vapour.
b The oceans, rivers and lakes are heated by the sun.
c The water falls as rain or snow and starts the cycle all over again.
d The vapour rises, then cools and turns back into water.

Chapter 3

The food connection

1 Do you like healthy food? Look at these pictures. What's healthy? What isn't? Find five examples of healthy food and five examples of not so healthy food.

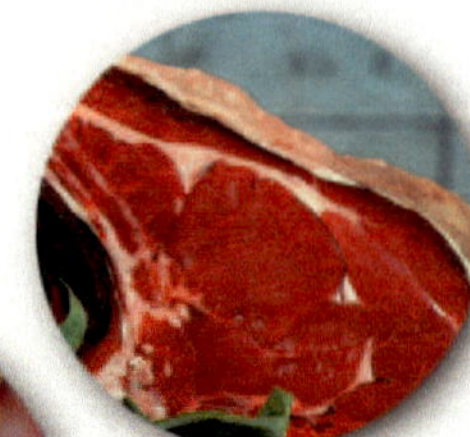

2 Make a list of food that keeps you fit and food that's fattening. Now mark what you like and/or often eat (✓) and what you don't like and/or don't eat very often (✗). Explain your list to a partner.

3 Think about how you usually feel before you eat the food on your list. Do you always eat sweets when you're sad, for example? Give two examples.

4 Write down your three favourite kinds of food. Now say what you think of when you eat them. Do you think of Italy when you eat spaghetti, for example, or do you think of a certain person when you see yoghurt?

5 Is there anything that you can't eat or drink? Explain.

6 Breakfast is important. What do you need for a healthy breakfast? Make a shopping list.

Helpful vocabulary

carrots ['kærəts] *Karotten, Möhren*
muesli ['mju:zli] *Müsli*
rice [raɪs] *Reis*
spinach ['spɪnɪtʃ] *Spinat*
yoghurt ['jɒgət] *Joghurt*

Helpful expressions

- I often/sometimes/never eat …
- Cornflakes are healthy / good for your health / keep you fit / …
- A lot of / Too much chocolate isn't very healthy / is fattening / can make you ill / …
- When I'm sad/happy/tired/hungry/…, I usually eat …
- When I see a commercial on TV / a model / a poster / …, I feel … / I start eating … / …
- Spaghetti / … reminds me of … / …
- I can't eat nuts/… because I'm allergic to them / because of my religion / …

MATERIALIEN FÜR

BILINGUALE KLASSEN

There & Then

NEW EDITION

Arbeitsheft für den erweiterten Englischunterricht Klasse 6

Vorbereitung für die Sachfächer Erdkunde und Geschichte

Cornelsen

MATERIALIEN FÜR
BILINGUALE KLASSEN

There and Then · *New Edition*

Arbeitsheft für den erweiterten Englischunterricht Klasse 6
Vorbereitung für die Sachfächer Erdkunde und Geschichte

Im Auftrag des Verlags erarbeitet von: Wolfgang Biederstädt, Köln, und Mervyn Whittaker, Bad Dürkheim

Beratende Mitwirkung: Mike Rooney, Nümbrecht, und Manfred Wildhage, Braunschweig

Redaktion: Birgit Herrmann, Aachen, sowie Kirsten Bleck und Katie La Storia (Verlagsredaktion)

Redaktionsassistenz: Bettina Schaschke

Illustration: David Barnett, Oadby, Leicester; Dr. Volkhard Binder, Berlin; Robert Broomfield, Tunbridge Wells, Christina de la Mare, Berlin; Gabriele Heinisch, Berlin; Dr. Hans-Joachim Kämmer, Berlin; Skip G. Langkafel, Berlin; Linda Rogers Associates, London: Jim Eldridge, Gary Rees

Gestaltung und technische Umsetzung: Susanne Meyer sowie Jutta Debo

Bild- und Textquellen: siehe Verzeichnis auf Seite 56

Contents

www.cornelsen.de

1. Auflage, 4. Druck 2007

Alle Drucke dieser Auflage sind inhaltlich unverändert und können im Unterricht nebeneinander verwendet werden.

Druck: H. Heenemann, Berlin

ISBN 978-3-464-31011-3

Inhalt gedruckt auf säurefreiem Papier aus nachhaltiger Forstwirtschaft.

Europe

Chapter 1

Europe is more than the EU (European Union). Not every country in Europe uses the euro, for example.

1 Look at the map of Europe on page 4. How many European countries are there?

2 Now look at these pictures (1–6) of famous monuments in Europe. What are their names? In which countries are they? Find the countries on the map of Europe on page 4.

1

3

4

2

5

Every country in Europe has its own symbol – an old monument, for example. But is there a symbol for Europe? What could it be?

6

3 Which of the monuments (pictures 1–6) symbolizes Europe most to you? You can pick two.

4 Find out which monuments your classmates picked. What are the two class favourites?

5 Now look at the box on the right. Find out which monuments people in different countries picked. Discuss your answers with a partner. Look at the "Helpful expressions", too.

Question: "Which two monuments symbolize Europe most to you?"

(in % = per cent)	England	Germany	Italy	Japan	Spain	USA
Eiffel Tower	62	39	50	55	59	72
Acropolis	14	12	20	10	11	12
Big Ben	49	12	13	7	14	46
Alhambra	2	5	3	6	36	1
Tower of Pisa	20	16	31	24	24	31
Brandenburg Gate	16	57	8	4	8	4

Helpful expressions

- Most people in … / (Only) …% of the people in … think/believe …
- (The) … symbolizes Europe most / is a European symbol / stands for Europe.
- I'm (not) surprised that … / It's interesting that …

6 Most people think the Eiffel Tower is a symbol for Europe. Find information about the Eiffel Tower – or another famous place in Paris. Now explain "your" place to a partner.
Helpful websites: http://www.paris-tourism.com and http://www.parisdigest.com

Chapter 1

Country	Adjective	Country	Adjective
Austria	Austrian	Italy	Italian
Belgium	Belgian	The Netherlands	Dutch
Bosnia-Herzegovina	Bosnian	Poland	Polish
(Great) Britain	British	Portugal	Portuguese
Croatia	Croatian	Russia	Russian
Czech Republic	Czech	Slovakia	Slovak
Denmark	Danish	Slovenia	Slovenian
France	French	Spain	Spanish
Germany	German	Sweden	Swedish
Greece	Greek	Switzerland	Swiss
Hungary	Hungarian	Turkey	Turkish
Ireland	Irish	Yugoslavia	Yugoslav

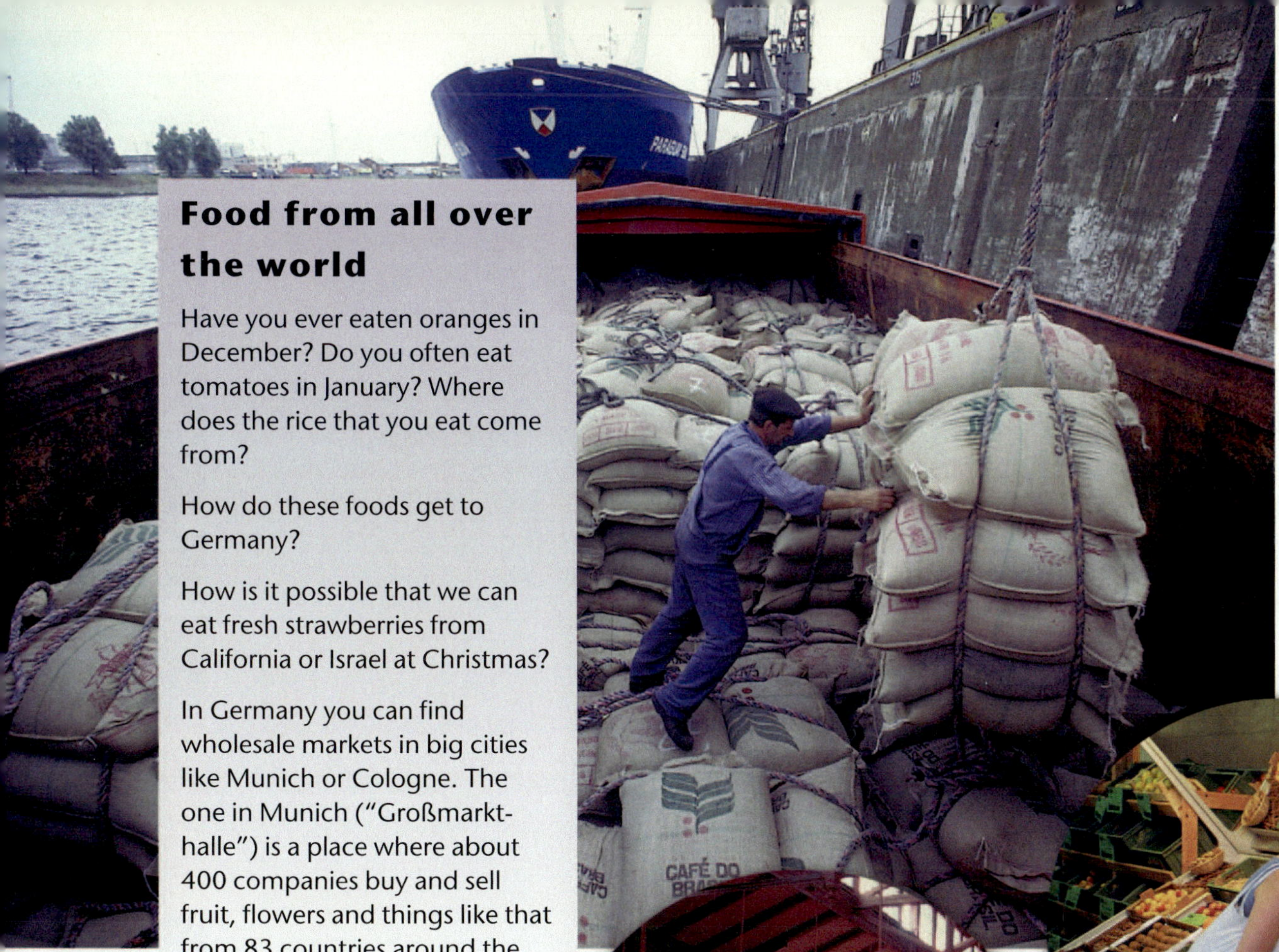

Food from all over the world

Have you ever eaten oranges in December? Do you often eat tomatoes in January? Where does the rice that you eat come from?

How do these foods get to Germany?

How is it possible that we can eat fresh strawberries from California or Israel at Christmas?

In Germany you can find wholesale markets in big cities like Munich or Cologne. The one in Munich ("Großmarkt-halle") is a place where about 400 companies buy and sell fruit, flowers and things like that from 83 countries around the world. The goods that are sold in a year weigh about one million tonnes. About five million people in and around Munich get fruit from here.

7 Go into a supermarket and find 5–10 products that aren't grown, produced or made in Germany. Find out where they're from. Then look at the map of the world on page 18 and try to find them there. If you can't find them, add them on a copy of the map.

8 Not only food ... Find out where the following products come from: cotton, rubber, wool.

Helpful expressions

You can say: *People in Brazil produce coffee.* But it sounds better if you use the passive: *Coffee is produced in Brazil.* Look at these sentences:

- Coffee/... is produced/planted/... in Brazil/...
- Bananas/... are grown/... in Costa Rica / ...
- The goods are packed into boxes/containers/...
- Then they're shipped/flown/exported/... to Hamburg/...
- They're taken by train/... to wholesale markets / supermarkets / ... all over Germany.
- They're sold to companies/customers/...
- People eat/drink oranges/coffee/... that are/is imported from Spain/Brazil/...

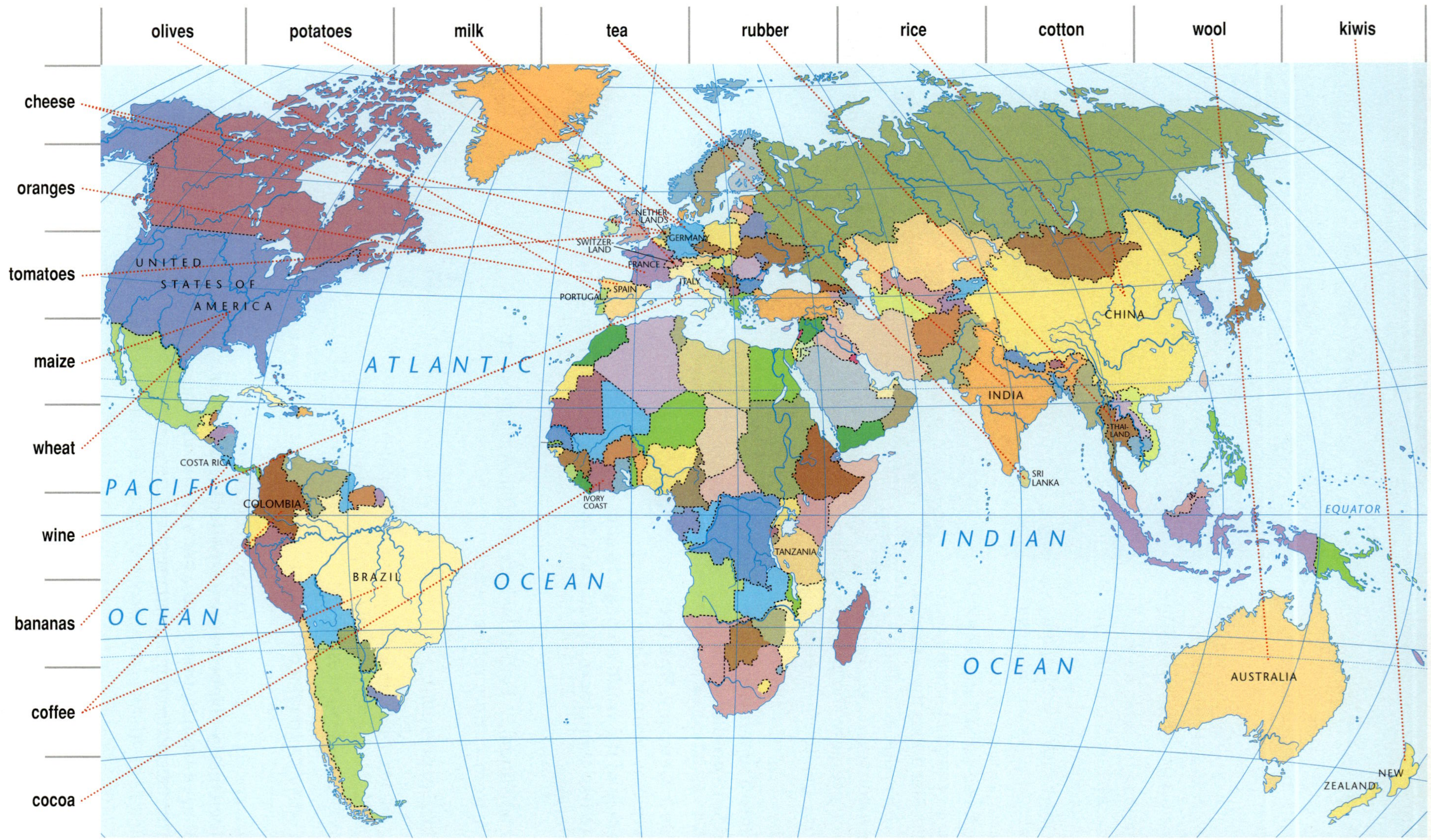
olives
potatoes
milk
tea
rubber
rice
cotton
wool
kiwis
cheese
oranges
tomatoes
maize
wheat
wine
bananas
coffee
cocoa
UNITED STATES OF AMERICA
COSTA RICA
COLOMBIA
BRAZIL
PACIFIC OCEAN
ATLANTIC OCEAN
INDIAN OCEAN
NETHER-LANDS
GERMANY
SWITZER-LAND
FRANCE
ITALY
SPAIN
PORTUGAL
IVORY COAST
TANZANIA
INDIA
SRI LANKA
CHINA
THAI-LAND
EQUATOR
AUSTRALIA
NEW ZEALAND

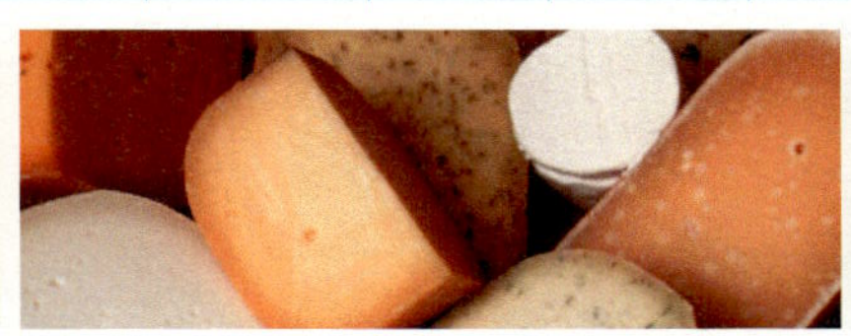

Pimhill Farm

Pimhill Farm is an old, but unusual farm in North Shropshire. It was one of the first organic farms in England. Organic farms don't use artificial fertilizers or pesticides and they keep the animals very carefully. This means that their products are a little more expensive than products from other farms. But some people think that organic food is healthier, so they pay more for it. Pimhill Farm is about 311 ha (= hectares), with 175 dairy cows. The milk is made into yoghurt and ice-cream. The main crops are oats, wheat and field vegetables. Silage and beans are grown for the cows. The farm even has a mill to make its own flour. It also offers an interesting service: People from the area can order boxes of farm products which are brought to their homes. This means that the food is as fresh as possible and doesn't need much packaging. Ginny Mayall is the owner of the farm. She has 10 full-time workers, 15 part-time workers and about 8 people for weeding and vegetable picking. Organic farms need more people because a lot of the work is done by hand.

1

2

3

9 Match the photos (1–3) to these sentences:

a A mill like this one grinds the wheat to make flour.
b Tomatoes and other vegetables are often grown in plastic shelters.
c The calves are kept in warm and dry hutches, but they can also go outside.

10 Look at the calendar below. Write down what it tells you about the farm year.

Crops	Jan	Feb	Mar	Apr	May	Jun	Jul	Aug	Sep	Oct	Nov	Dec
Potatoes			Planting	Weeding				Early potatoes	Harvesting main crop			
Beans			Planting						Harvesting	Storing as feed		
Vegetables	Planting and cropping											
Arable	Weeding (by machine)							Harvesting	Ploughing & planting			
Grass					Silage making	Cutting grass to prevent weeds		Silage making				
Cows				Cows go out into fields						Cows come in		
					Silage making		Silage making					
	Dairy cows are milked and have calves											

Helpful expressions

• In March	potatoes	are	planted.
• In June and July	the grass	is	cut.
• Every day	…	are	milked.
• …	…	come in	from the fields.

11 Find out:

a where Pimhill Farm is. (Use a map of Britain.)
b the differences between organic and other farms.
c other unusual things about this farm and its products.

Helpful website: http://www.pimhillorganic.co.uk
Helpful address: Pimhill Farm, Harmer Hill, Shrewsbury, Shropshire SY4 3DY

Coffee from Tanzania

European settlers started growing coffee on the slopes of Mount Kilimanjaro in Tanzania a hundred years ago. In the 1960s coffee became important for Tanzania's economy. Today over 90% of Tanzanian coffee is grown on small farms like Antippi Estate. Antippi is about 47 ha and it has 6,000 old and 750,000 new coffee trees. 1,250 coffee trees are planted per acre. They produce a full crop after three years – 1 to 1.5 kilos of coffee beans. The farm also has banana trees, lemon grass and maize. It employs between 30 and 80 workers on a daily basis.

Antippi Estate

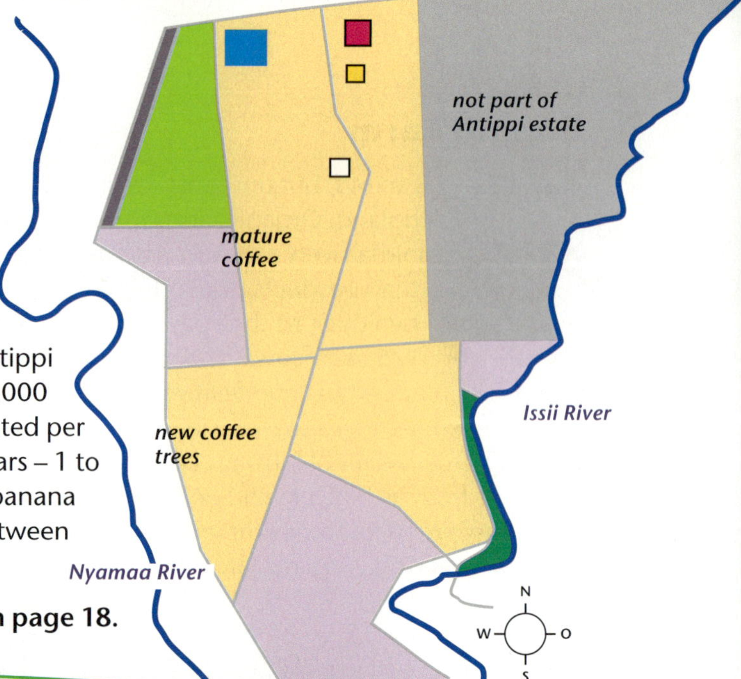

12 Find Tanzania on the map of the world on page 18. Where in Africa is it?

13 Now look at the map on the right. What kinds of coffee fields can you see? What other crops are grown? Where do the coffee workers get water from? Write down everything that the map tells you.

Pulping coffee

Weeding coffee

Planting coffee

Helpful expressions

- … is the (second) most important product/crop/…
- … is (also) grown on the farm.
- … is/are next to / between / …
- … is/are planted between … and …
- Water is collected in …
- The … house(s) is/are in the middle of / near / …

Coffee throughout the year
Coffee trees are planted by hand during the two rainy seasons. They must be watered very often during the dry season from September to November. The weeding is done by hand or by a small tractor. The main harvest is from June to September. The coffee beans are picked by hand and brought to the farm factory. There they are pulped to take away the soft outer layer from the bean. They are fermented in water for 2–3 days, then washed and dried for 10–11 days. After that the coffee beans are sold to exporters in Moshi, a small town 11 km south of the farm. They sell it as "Columbian Mild" coffee to Europe, the USA and Japan.

14 Read the text on the left. Draw a timeline for one year and mark what is done to grow coffee throughout the year.

15 Compare Antippi farm with Pimhill Farm (page 19). Make a list. Look at the size, the number of workers, the crops, machines and animals, for example.

16 Compare the climate of Mwanza, Tanzania (Antippi) with the climate of North Shropshire, England (Pimhill).
Helpful website:
http://www.klimadiagramme.de/Afrika/mwanza.html

Recipe: Chocolate Pudding Cake

(9 people; 45 minutes)

Everyone in your family will like this special dessert. Serve it warm with a bit of whipped cream and a tall glass of milk.

You need:

Batter		
1	cup	flour
1/2	cup	sugar
1/2	cup	milk
2	tablespoons	unsweetened cocoa
2	teaspoons	baking powder
1/4	teaspoon	salt
1 1/2	tablespoons	vegetable oil
1 1/2	teaspoons	vanilla

Topping		
1/2	cup	brown sugar
3	tablespoons	unsweetened cocoa
1 3/4	cups	boiling water

a tablespoon

a teaspoon

Then:

Batter: Put everything in a mixing bowl and stir until smooth. Spread batter evenly in a 9 x 9-inch baking tin.

Topping: Put brown sugar and cocoa in a small mixing bowl and stir until blended. Break up lumps of brown sugar with a fork.

Sprinkle the topping evenly over the batter. Pour boiling water over everything. **Do not stir!** Put in oven (180°C) for 35 minutes.

▸ p. 55

Cooking project

Collect recipes for healthy meals. If possible you can try and cook one of them for your class.

Food safety facts

Cooking is fun, but: Every year millions of people become ill – and they don't know that it was food that caused their problem. Below are some facts and tips about food safety. Make sure that you and your family keep to these rules!

Wash your hands!

Always wash your hands well, with soap and warm water, before and after handling food. Our hands often carry bacteria on them. If we transfer those bacteria to food, it's a good place for them to grow! But it's also important not to let the bacteria from raw food stay on your hands from where they could get to your mouth or to other food.

Keep hot food hot!

If food is cooked, make sure that you keep it hot if you don't eat it right away. Bacteria could make you ill.

Keep cold food cold!

Cold salads, dairy products and other food which need refrigeration should always be kept below 4°C. If they become warm, bacteria can grow fast and become dangerous.

Wash the food!

Because fresh fruits and vegetables are grown outside, they can come in contact with a lot of bacteria. Most of these bacteria aren't dangerous, but it's important to wash fresh fruits and vegetables under running water before you eat them.

17 Which of the above safety facts do you think are most important?

18 Find more safety facts. If you need help, ask your mother or father. Then find some attractive pictures and make a big poster with all your safety rules.

Chapter 4

Black Death

1 Look at the picture below. What do you think is happening in it?
These words can help you: angel – devil – bow – arrow – (to) shoot – (to) hit – (to) pour

2 Find out what the "Black Death" is. What has the picture above got to do with the Black Death, do you think?

3 Before you read about the Black Death (1347–1352), collect words and phrases about illnesses in a network. Look at the "Helpful vocabulary" on the right, too. You can start like this:

Helpful vocabulary
armpit ['ɑːmpɪt] *Achselhöhle*
boil [bɔɪl] *Furunkel* • brave [breɪv] *tapfer*
cold [kəʊld] *Erkältung* • fever ['fiːvə] *Fieber*
(to) groan [grəʊn] *stöhnen*
groin [grɔɪn] *Leistengegend*
illness ['ɪlnəs] *Krankheit*
pain [peɪn] *Schmerz*
pus [pʌs] *Eiter* • smell [smel] *Geruch*

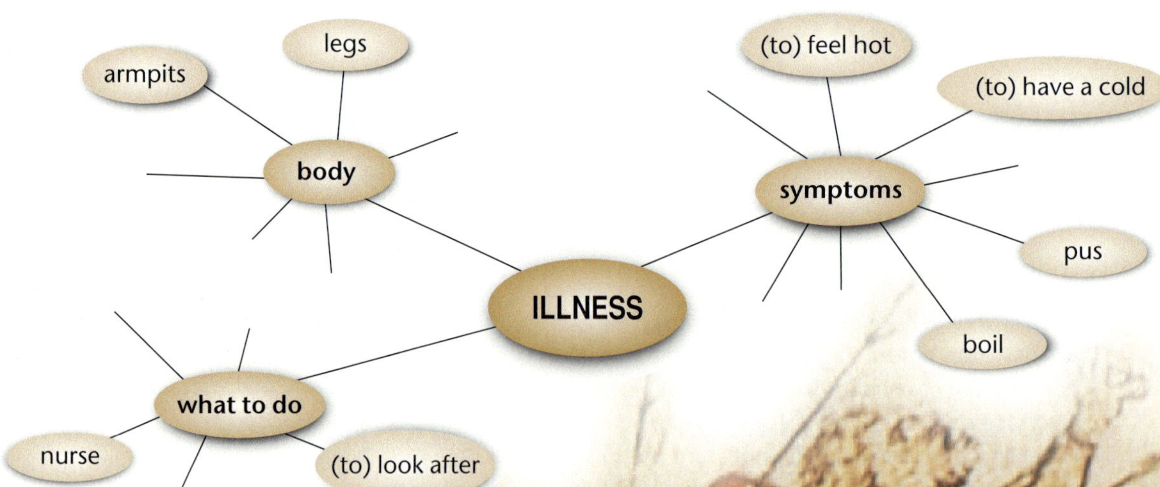

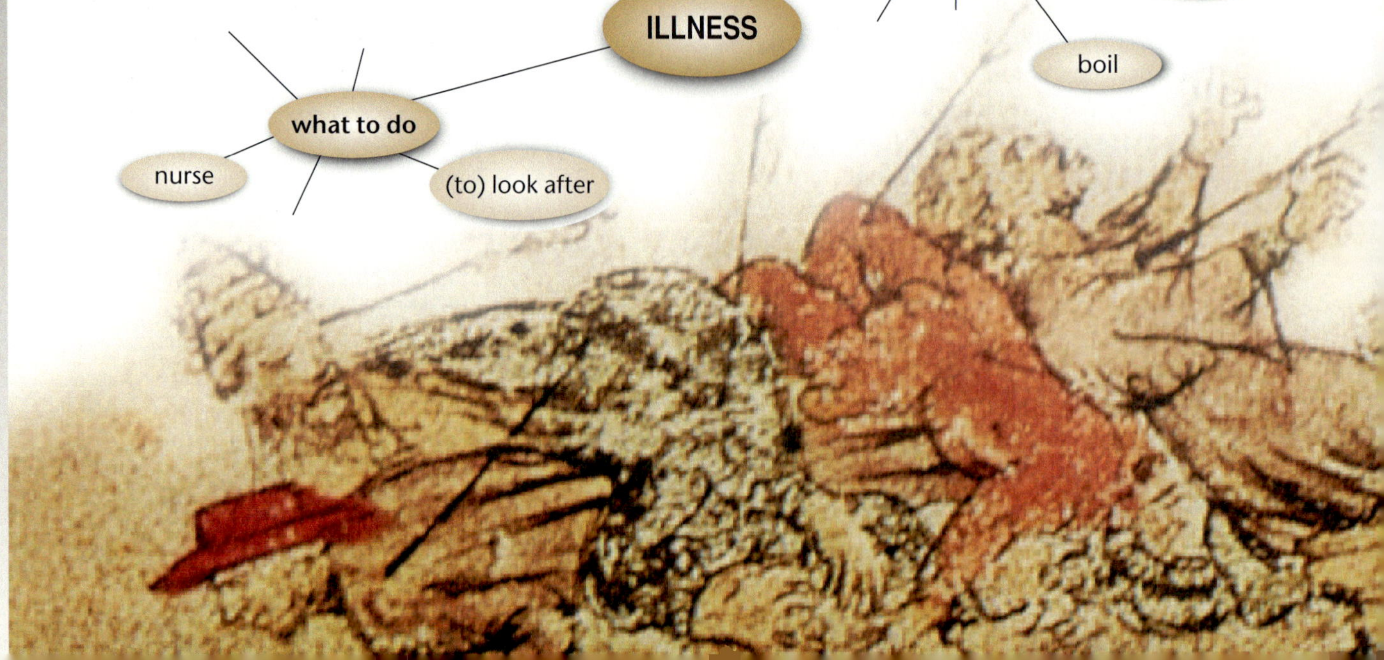

Alice's Wedding Day

4 Before you read the text, look at the pictures. What do you think the story is about?

5 In this story you're going to see tenses that are new to you. For example:
They had met four months ago … (l. 7)
He was selling things … (l. 10)
I'll (= I will) find a nice place … (l. 26)

Look at the context and say what these sentences mean.
Then say what tense you use in German.

The year is 1349. Alice lives in the village of Wharram Percy in Yorkshire, England. She's 25 years old and today is her wedding day. She's marrying Stephen, who drives packhorses for a merchant …

Alice woke up and heard the church bells. It was a beautiful sunny morning, and a perfect day for a wedding. Alice was going to marry Stephen today. Stephen drove packhorses for a merchant in Southampton.

They had met four months ago, in April. Stephen had stopped at her village, Wharram Percy, on his way to York. He was selling things for the merchant and Alice had chosen some red ribbon. Alice remembered how he had smiled at her, and she had smiled back.

A week later, Stephen was there again, on his way back from York. This time, he stayed a whole month. At the end of the month, he had asked Alice to marry him. Alice wasn't sure; she would have to leave home and go to Southampton with him. But her mother said, 'You must go. Stephen is a good man.' So Alice said yes, and the wedding day was set for 15th August.

Stephen went back to Southampton. 'I'll find a nice place for us to live,' he said when he went. 'And I'll bring you a nice wedding dress, too!' he smiled.

Alice met Stephen when he and his packhorses stopped in Wharram Percy on their way to York.

And now the wedding was only a few hours away. Alice was excited. She thought back to last night, when Stephen had arrived.

How lovely it had been to see him again. He arrived very late. Alice was waiting with her parents and her best friend Sarah. Nearly everyone in the village was asleep. Stephen was cold, tired and hungry. He had been travelling for two whole weeks and told stories of a strange and terrible new illness at home in Southampton.

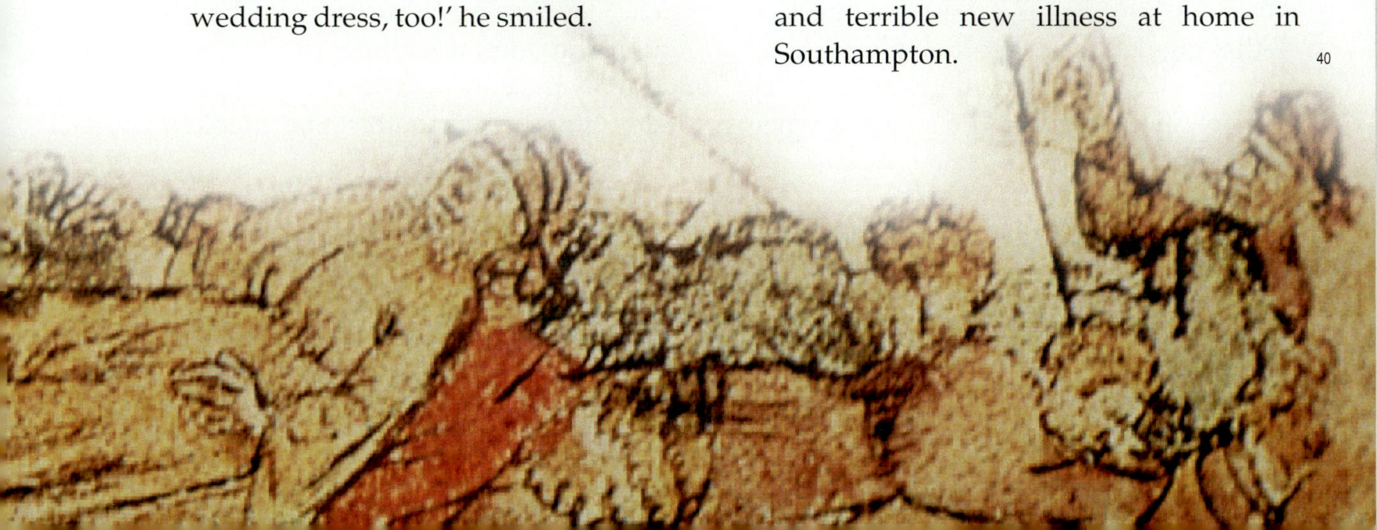

Chapter 4

Stephen arrived from Southampton. He said people there were dying of a terrible illness.

Alice said to Sarah, 'He's so tired that he's talking nonsense!'

'People in Southampton are dying of boils that are as big as eggs,' Stephen said. 'When the boils break, the smell is so bad that even the dogs and rats die. The people drink and drink, but they can never get enough water.'

Her parents laughed at this.

Stephen got angry because people were laughing at him.

But he felt better when Alice gave him some hot soup and some cheese. He went out to his packhorse and carried a big bag into the house. In the bag was the most beautiful dress that Alice had ever seen. It was cream-coloured and gold with red patterns to match the ribbons that Alice had chosen when they first met. It was Stephen's wedding present for her.

Alice shook the dress out. Some fleas jumped out and her father swore when they bit him. But the dress was perfect. Alice giggled when her friend Sarah tried it on first. Then her mother wanted to try it, too. Alice could not stop laughing when her mother tried to squeeze into her wedding dress.

'You'll squash all the fleas in there!' her father said.

Stephen had brought other things for the wedding, too. There was some ham, a whole cheese and then something really special: salt, cinnamon, nutmeg and pepper. 'Stephen must have some rich friends in Southampton if he can get these spices!' thought Alice.

But Alice worried when she remembered how Stephen had found a big black rat at the bottom of his bag. It was dead and Stephen threw it on the fire before Alice's mother could see it.

Stephen found a dead rat in the bag which he had brought with him from Southampton.

She said that seeing a dead rat always brought bad luck to a family.

'But today there'll be no bad luck,' Alice thought as she got out of bed. 'Today we'll get married. The whole village will come to the wedding.' Then she heard her mother. 'Come on, Alice, get up. We must fetch Stephen from cousin Elizabeth's house. The priest is waiting.'

Suddenly the door was thrown open, and there stood Elizabeth. She looked frightened.

'Oh dear, what's wrong?' asked Alice.

Stephen was too ill to go to the wedding.
He had a fever.

'Come with me,' Elizabeth answered. 'Stephen is ill. He has a fever and he's asking for you!'

Alice ran out of the house and across the way to Elizabeth's home. Stephen was still in bed. He groaned and lifted his head.

'Don't worry, Alice, it's only a cold,' he said. 'I'll be better tomorrow. Tell the priest to wait. We can get married on Monday.'

In the evening Stephen got worse. He had terrible pains in his armpits and groin. And there were lots of black spots on his arms and legs. What should they do?

Alice was brave enough for all of them. 'Elizabeth,' she said, 'you must go and stay with mother and father. I'll stay here with Stephen. Ask the wise woman for some leaves and flowers to make tea. Leave all the food, the herbs and the drinks at the door. I'll come out to get them. It'll be better if only I touch Stephen from now on.'

With those words, Alice left her parents and cousin and went inside the house to look after Stephen.

Monday came and there was no wedding. Stephen was even worse. For four days he lay groaning and crying on his bed. 'Don't touch me!' he cried. 'God has sent his devils to punish me because I drank beer on Sundays. Give me water. God has sent this terrible thirst to punish me.'

On the fifth day, Stephen looked at Alice. 'Pray to God that I may go to heaven. Call the priest! Call the priest! I'm dying. I must make my last confession.'

But it was too late. Stephen's eyes rolled back into his head. His body moved again and then all was still. Alice lifted the blanket to pull it over his face. When she lifted it, the most terrible smell hit her. Poor Stephen. Under his arms and in his groin

were the biggest boils that Alice had ever seen. They had broken and were oozing a thick, black, slimy pus. That was where the smell came from.

Alice was so brave. She wrapped Stephen's body in his blanket. Then she took her blanket. She wrapped Stephen's body in this, too. Now, she put some more wood on the fire and collected everything together; the food, clothes and the rest of the bed. She threw everything on the fire. She put some sweet-smelling herbs in, too. Then she took a last look at Stephen in his blanket and slowly left the room to tell her parents the bad news.

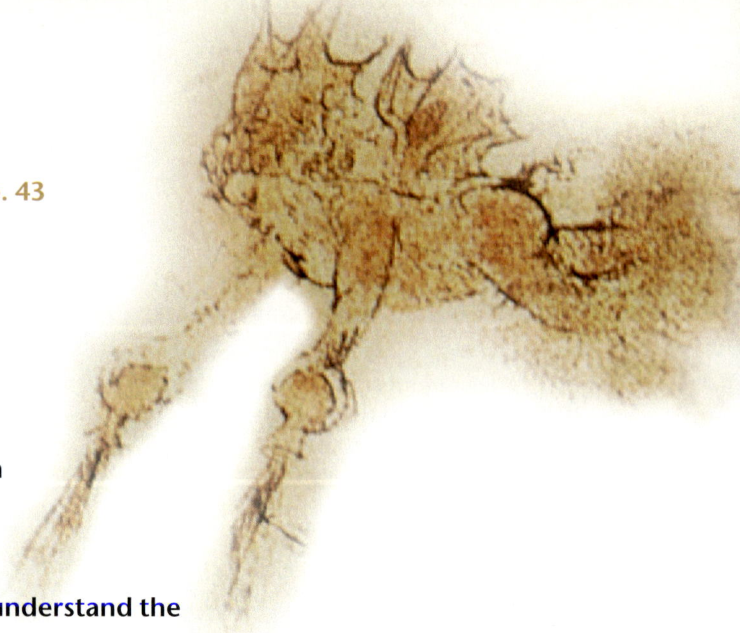

Thinking about the story ▸ p. 43

6 Ask and answer the *wh*-questions *(who, where, when, what, why)* and *how*.

7 Make two lists:
(1) everything the text tells you about the "Southampton illness";
(2) everything, good and bad, that Stephen brings with him from Southampton.

8 **Look through the text again and find people's reactions to the illness. Do they understand the Black Death? Explain your answer.**

9 **Some historians think that the Black Death was two deadly plagues that attacked Europe at the same time. Read this information about the two different illnesses:**

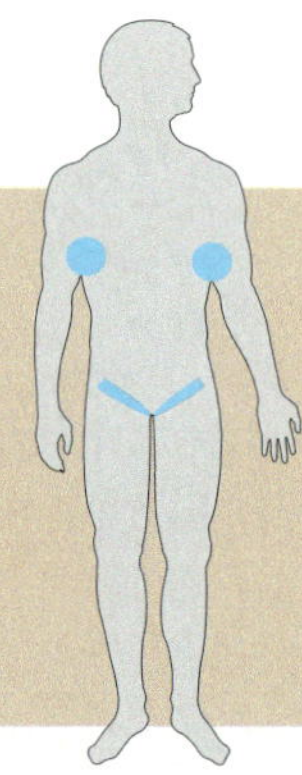

Bubonic plague
Rats carry this plague. When fleas bite the rats, they can get infected. The fleas then bite people and give them the plague. People with this illness have a fever and big boils in the groin and armpits. About 70% of people who catch it die. This usually takes between four and seven days.

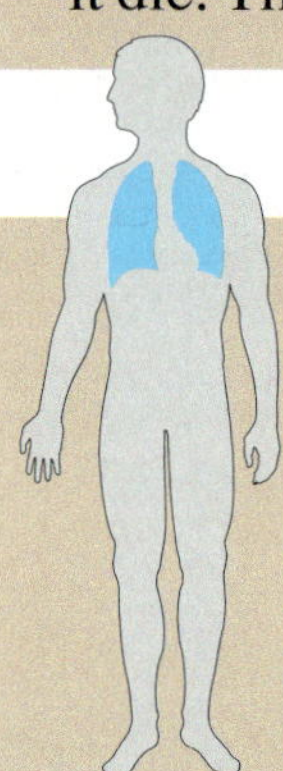

Pneumonic plague
This plague is caught through breathing. It attacks the lungs. People with this illness cough blood and infect people every time they breathe out. Another sign of pneumonic plague is a fever.
Everyone who catches this plague dies after about two days.

Which illness do you think Stephen had? Explain your answer.

10 **Do you think Alice was a good nurse? Why/Why not?**

11 **What do you think happened next in the village of Wharram Percy? Work with a partner and tell the rest of the story. Perhaps you could write a dialogue between two people from the village who didn't die.**

12 Find out more about the Black Death. **How did people try to stop it? What happened when lots of people in one town or village caught it? How many people died? Can you find some pictures from the time? (A library or the Internet can help you.)**

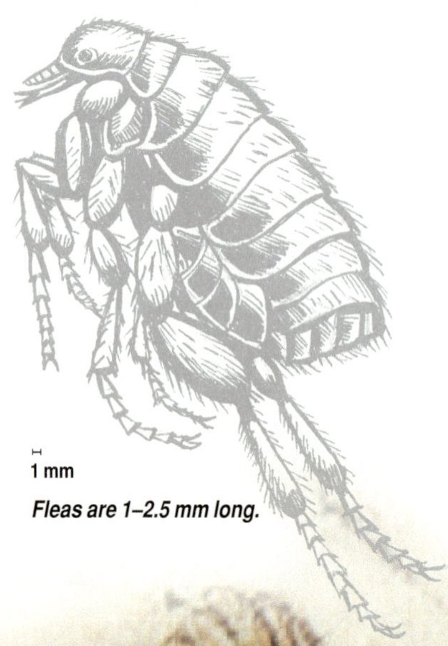

Fleas are 1–2.5 mm long.

Some special days

31 October

Hallowe'en is really two words: *hallow* (= German "heilig") and *e'en* (= short for *evening*). A long time ago this was the beginning of the New Year. People thought the dead left their graves, so they put on frightening masks and costumes to scare them away. Today you can see **ghosts, bats, witches with their cauldrons, were-wolves, Frankensteins** and **Draculas** walking with their **jack-o-lanterns** through the night on 31st October – on their way to a **Hallowe'en party.**

1 Can you match the words above to the pictures below?

A SCARY STORY

2 Read the beginning of the scary story on the right.
3 Now think about how the story could go on. Copy the beginning and write the ending. Remember to use the past tense.

Helpful questions

- Who is the little voice? Is it something nice or nasty?
- What did Lou do next?
- Think of a good ending. Can you surprise the reader?

Not alone

It was a cold, windy October night. Lou's parents had gone out for the evening and he was alone in the house.

He went to bed early. He looked in the cupboard, behind the curtains and under the bed. Then he locked the door and turned his bedside light off.

At that moment he heard a little voice. It said, "Oh good, now we're all alone together …"

1 January

"Out with the old, bring in the new" is the motto of the **New Year**. Some people make **New Year's Resolutions** because they want to change.

If your teacher keeps the resolutions safe until next year, you'll see how much you've changed.

1. I'm going to be in Maths
2. I'll argue less my brother.
3. I'll try to save money for my ne

New Year's Resolutions

4 Think hard about what you want to do differently in the coming year.
5 Write 3–5 resolutions on a piece of paper and put it inside an envelope.
6 Close the envelope. Write your name across the back; no one can open it now. On the front, write New Year's Resolutions for 20… and your name and class. Then write Don't open before 1st January 20… .

Helpful questions

- Do you spend too much money?
- Should you be nicer to your friend/sister/…?
- Do you want to spend more time with your pets?
- Do you want to train harder for your team?
- How do you want to help at home?

Topic

14 February

14th February is **Valentine's Day**.
Young people in Britain get messages like: "Be my Valentine" or "You're so sweet", but the **cards** never have the sender's name on them – so you have to guess: Who sent the **card**? Who seems to be interested in me?

Roses are red,
Violets are blue,
Sugar is sweet,
And so are you!

Valentine's poems

Often the **cards** have little **poems** like these:

THINK OF M,
THINK OF E,
PUT THEM TOGETHER
– AND THINK
OF ME!

But usually the **poems** aren't really serious …

7 Look at the boxes below. The two lines in the boxes on the left are the beginning of a poem. Write them down. Then find two lines from the list on the right to finish the poems.

Think of me on a river,
Think of me on a lake,
…

Butter is butter,
Jam is jam,
…

I bet you can't guess
And send me a piece of cake.
Think of me on your wedding day
Who I am!

8 Now try to finish these two poems with your own words.

Do you love me,
Or do you not?
…

Roses are red,
My nose is blue,
…

1 April

It was on **1st April**, 1957 – **April Fool's Day**.
The BBC TV announcer said, "And now we're going over to Switzerland for a report on their spaghetti trees". A short film showed people working on a "spaghetti farm", cutting the "ripe" spaghetti that hung from the trees … The BBC's **joke** was so successful that viewers phoned to ask where they could buy a spaghetti plant!

April Fool's Day Project

Start to collect **jokes** for **1st April**. Put them on the classroom wall.

Helpful tip

Most jokes are funny in English *and* German, but sometimes it isn't easy to translate them because the joke is in the sound of the words. Use your dictionary carefully if you want to tell your jokes in English.

9 Look at the jokes below. Add one joke to each of these three groups: doctor jokes, school jokes, waiter jokes.

Waiter, your thumb is in my soup!
Don't worry, sir. It isn't hot.

Doctor, doctor, I've only got 59 seconds to live!
Wait a minute.

I can't go to school today. I'm sick.
Where do you feel sick, dear?
In school!

Remembering around the year

Here's a list of special dates from the past – one for each month of the year. Split into twelve groups. Each group chooses one month. Your task is to find out more about that month's special date. (You can go to a library or use the Internet, for example.) Collect notes and pictures, and prepare to give a two-minute talk: In the first week of "their" month, each group will tell the class what they found out.

JANUARY Queen Victoria of Britain died on 22nd January, 1901. How long did she rule? Who was her husband? Find out about her life and times.

FEBRUARY On 11th February, 1990, Nelson Mandela was released from prison – after how many years? Why was he in prison? Tell his story.

MARCH Alexander Graham Bell invented something important on 10th March, 1876. You probably use it every day. What was it? Find out details.

APRIL The Titanic sank in the Atlantic Ocean on 14th April, 1912. Why? What's the story behind this disaster?

MAY The Channel Tunnel was opened on 6th May, 1994. Where is it? What can you find out about it?

JUNE Queen Elizabeth II was crowned on 2nd June, 1953. Is she still the queen of Britain? Who are her children and grand-children? Tell the class about the Royal Family.

JULY The Scouts Movement started on 31st July, 1907. What are the scouts called in German? What do they do? Find out more about them.

AUGUST Alfred Hitchcock was born on 13th August, 1899. What is he famous for? Find out more.

SEPTEMBER The American Mark Spitz won a gold medal on 4th September, 1972. It wasn't his first. How many medals did he win? What did he win them for? Find out details.

OCTOBER Mahatma Gandhi was born on 2nd October, 1869. He helped a country to become independent from Britain. What country? What happened? Find out.

NOVEMBER The Canadian James Naismith was born on 6th November, 1861. What popular sports game did he invent? Tell the class more about it.

DECEMBER On 8th December, 1980, John Lennon was killed in New York. Who was he? Where was he born? Find out more.

Chapter 5

Survival!

Survival means staying alive in difficult conditions; after an accident, in extreme climates or difficult landscapes, or when you're alone. Books like the ones on the right are about survival.

1 You've just crash-landed your plane in the desert, at a lonely place far away from people. You wait a day, then you decide to go and look for help. First you look through the plane and you find:

lighter *mirror* *chocolate* *blanket* *snakebite serum* *knife* *water*

flares *diary* *sunscreen* *string* *pistol* *shovel* *sunhat*

a But you can take only five things with you. Decide which five you want and write down why they would be useful.
b Now listen to your partner's ideas. Tell him/her what you think of his/her choices.
c Together try and decide on a new list of the five most important things for the journey.

Helpful expressions

- You need … because …
- … is/are useful for …
- … is/are good when …
- Yes, that's a good idea.
- Well, you can't do much with …
- You don't need … when … Water/… is/are much more important.

Getting water from the desert

In task 1 on page 30 most of you probably decided to take water because it's the first thing that you need. Here's an idea from a survival handbook. It shows how you can get water from the ground, even when you're in a very dry place.

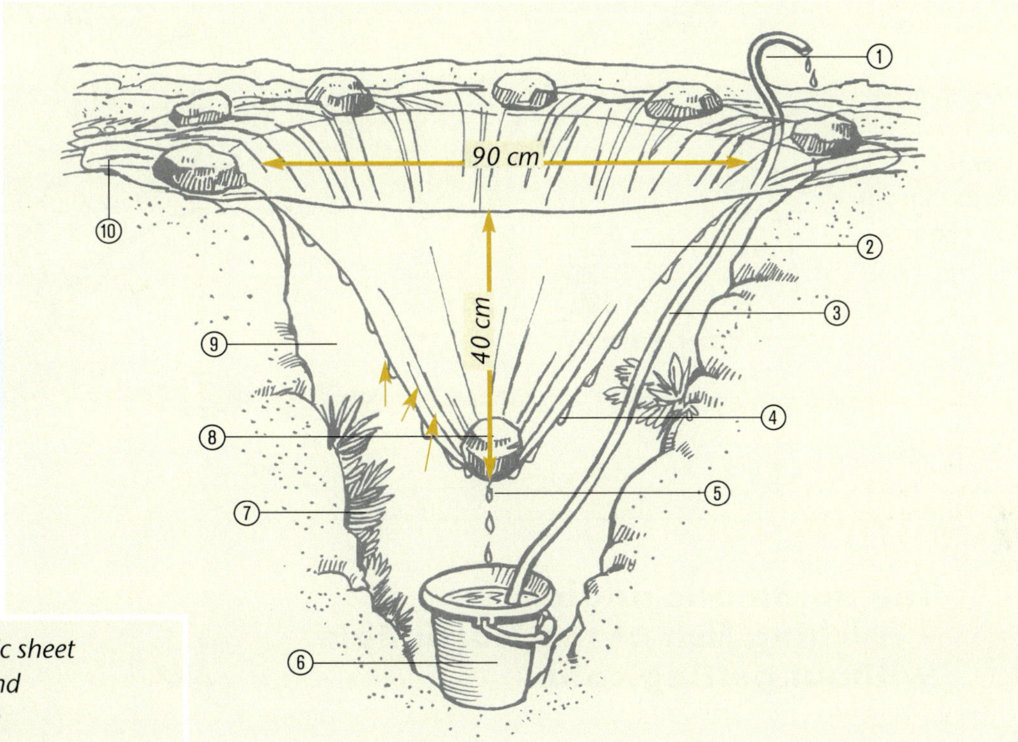

a plastic sheet
b ground
c stone
d bucket
e plastic pipe
f extra plants
g water evaporates
h moisture condenses
i water collects
j water is sucked up

2 On a copy of the drawing write down the words from the box (a–j) next to the correct numbers (1–10).

3 Use the labelled drawing and explain how the construction works.

4 Now you can write the text for the handbook. Write it under these headings:
a What you need.
b How you make it. *Useful words: (to) dig a hole; (to) weigh down.*
c How it works. *Useful words: (to) condense; (to) collect; (to) suck the water up.*

The handbook says that snakes and other animals sometimes smell the water and come to drink. But they fall in the bucket and can't get out again. Then, the handbook says, you have something to eat, too!

Survival at the end of the world

At the poles, there's enough water. But what about food? You can't eat snow, so you have to catch fish. But it's too cold to stand next to a hole in the ice all day. You need – the automatic angler! Read here how you can make one. Look up words that you don't know.

The automatic angler – catching fish at the North Pole without getting cold

You need two sticks (about 1 m long), some string, a plastic bag, some fishing line (about 3 m long), a hook and bait.

Lay the two sticks in the form of a cross with one long end and one short end. Tie them with string so that they are fixed together at right angles. Tie the plastic bag to the long end of the cross, so that it looks like a flag. Tie the fishing line, with the hook and bait, to the short end.

Now cut out a hole in the ice of about 50 cm diameter. Lay the crossed sticks across the middle of the hole. The flag is now lying on the ice outside the hole, the short end of the stick with the line is over the water. Drop the hook and line into the water.

When a fish bites, it pulls on the line. This pulls down the shorter end of the flag stick into the hole – and makes the flag flip up.

An angler can watch three or four fishing holes at the same time. Every time a flag goes up, the angler can go and collect the catch.

5 Now try and draw the construction. Follow these steps:

a First draw the hole in the ice, looking down on it.
b Now draw the automatic angler before a fish bites, again looking down.
c Then draw a new picture, from the side this time. Show the fish on the hook and show what happens on the ice when the fish pulls on the line.

Survival on Lost Island

6 This is a map of Lost Island. Imagine there are all sorts of dangers on this island that make survival there difficult. What dangers could there be? Copy the map and mark them on the copy. For more ideas look at page 34.

Here are some ideas for your map of Lost Island. Look at the map on the right that Carolin (13) drew and at the box below.

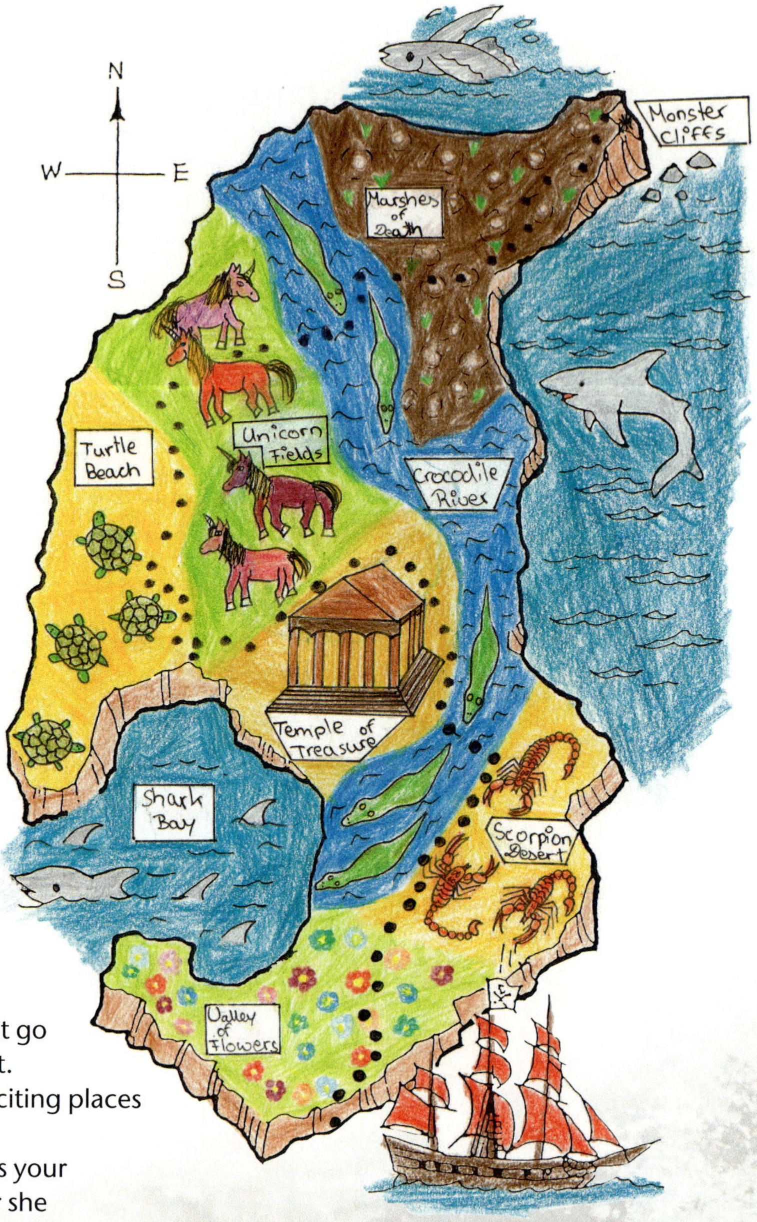

- Cannibals' Jungle
- Caves of Gold
- Field of Paradise
- Forest of Friends
- Ghost Village
- Hills of Holes
- Piranha Lake
- Snakes' Path
- Tigers' Track
- Vulture Mountains

7 When your map is finished, you must send your partner on a mission. He or she must start somewhere on the coast and find something important at a secret place in the middle of the island. This is what you have to do:

a Decide what the mission is, what your partner must find, where he or she must go to find it and where he or she must start.

b Decide on a route that crosses some exciting places on your map.

c Write a "message in a bottle" that sends your partner through the island so that he or she arrives at the secret place. Give him or her tips about how to survive the dangerous places. You can use some expressions from Tom's message below for your own message.

To get the golden key to the temple, you must be brave and strong. The island is dangerous – be careful!

First land your ship at Turtle Beach. The turtles will try to bite your legs. But throw them a magic apple and they'll let you pass.

Now turn left and go straight on. You must go north and climb up the Volcano of Terror. If it erupts, dig a hole quickly and hide in it.

When you've climbed down the other side, you

d Follow the route that your partner has written for you. (You could make a line in pencil.) Does it take you to the secret place? Where is it?

Survival holidays

For some people survival isn't a question of life or death, but an adventure and like a sport. There are special survival holidays, where you can be in different landscapes, meet some dangers, and – with the right equipment – do lots of exciting activities.

Antonia (12) has put together "survival adventure" words and expressions in this network:

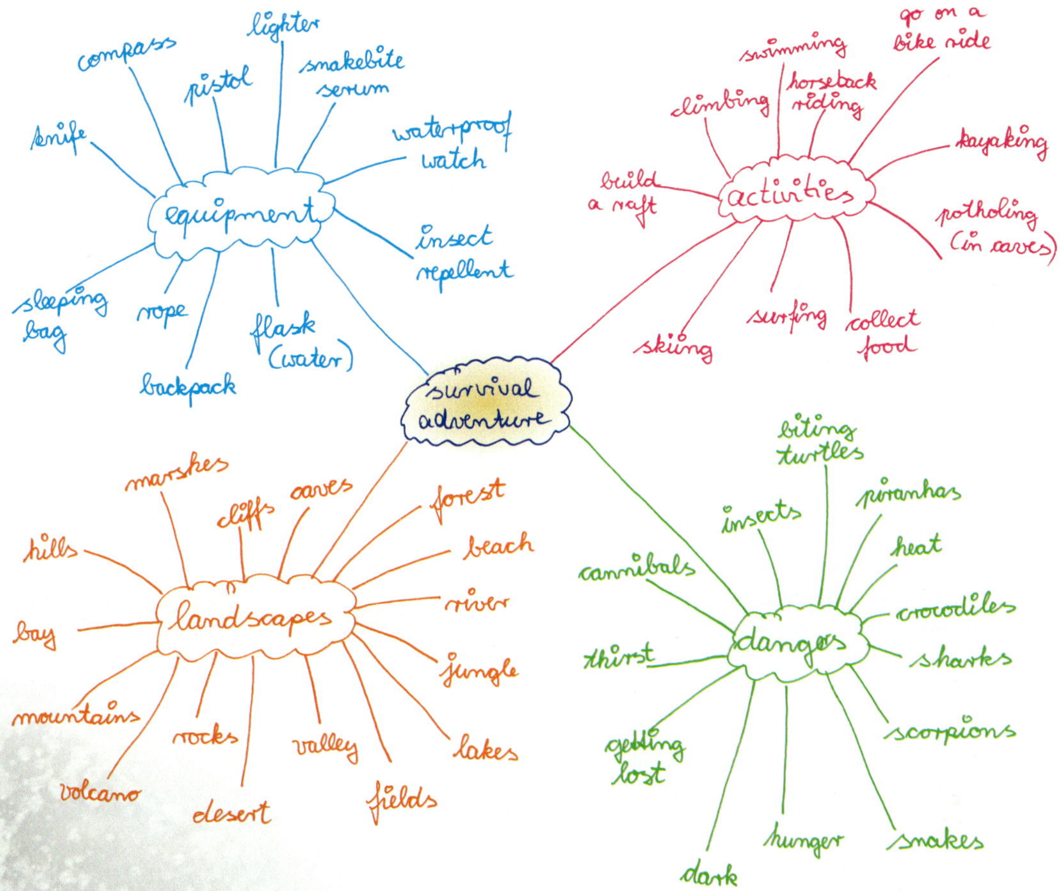

8 Imagine that you've bought Lost Island and now you want to offer survival adventure holidays to tourists. So you start to make a brochure.

a In the brochure you want to tell people about …
- the landscapes they can find on the island.
- the activities they can do.
- the equipment they should bring and the equipment you can give them.
- the dangers they can expect.

Look through the network and choose a few things from each "cloud".

b Now design the brochure.
- Plan and write the text about the exciting survival adventure holiday.
- Think of a motto or headline.
- Design the brochure – perhaps a small (A5) folded brochure. Use colours and maybe a computer layout to make it attractive. Remember, you want adventurers to come to your island.

c Put all your brochures on a table in the classroom and look at the ones that the others have made.

Chapter 6

The Norfolk Broads

For this year's "New Horizons" Mrs Hill is planning to visit the Norfolk Broads.

SEE THE NORFOLK BROADS

ORGANIZER Mrs J Hill
NUMBER OF PUPILS 20 COST £50

We're visiting the Norfolk Broads, Britain's only lowland, wetland national park. We want to visit places like Potter Heigham and Wroxham. We also want to find out why the Broads are in great danger.

1 A tour through the Broads

a Look at a map of Britain and find out how far Potter Heigham is from Chester.

b Imagine you want to visit the Norfolk Broads. Look at the map on the right, describe a round trip and say what you can see and do on your tour.

Helpful expressions

- From Norwich we go to / leave for …
- It's north-east/south/… of …
- … is situated east/… of … / near … / on the coast.
- The place west/… of … is called …
- The A47/… runs between … and …
- There's a nature reserve / … near …

▸ p. 41

2 Mrs Hill is showing her pupils the list of facts and figures on the Norfolk Broads below. Everybody has to write down two questions for the tour guide. What are your questions when you look at the figures?

FACTS AND FIGURES ON THE NORFOLK BROADS

Area	about 303 km^2
Population	5,500
Visitors	2 million per year
Private boats	10,640 (1995)
Boats to hire	2,445 (1995)
Land uses:	
fenland	8%
woodland	11.5%
grazing marshes	43%
farmland	27%
open water	6.5%
other	4%

Too many tourists

On the first morning of their trip to the Norfolk Broads Mrs Hill's group met a tour guide. Here's what he told the children about the Norfolk Broads.

GUIDE The Norfolk Broads is one of Britain's national parks. It's also the most famous lock-free waterway in Britain. There are 200 kilometres of rivers and shallow lakes. These lakes are called 'broads'. Long before the Broads became an important holiday area, the waterways were the only link between the villages here and the rest of the country. The land was marshy, so people couldn't build roads and bridges for many years. They carried coal, tea, sugar and farm products on the waterways in 'wherries'. That's a special kind of sailing boat. Today there's only one trading wherry left.

KATE So when did the Norfolk Broads became a famous holiday area?

GUIDE That was in the 1870s, when the railway came to Norfolk. Visitors could get here easily from many parts of England. The Broads became 'the place' for rich people. They took an adventure holiday here, and many of the wherries were used as holiday homes.

JOHN What about all the motor cruisers around here?

GUIDE They came in the 1920s. They meant that people could go on a Broads holiday, even if they didn't know how to sail. Today there are more than 5,000 motor boats. A lot of people are happy about this because the tourists bring money and jobs to the area. But they also bring problems.

MRS HILL Can you tell us something about the problems to the environment?

GUIDE Well, first of all, there's the water quality. Before 1900 the Broads had clear water with a lot of fish and wildlife. From about the 1950s the rivers and broads began to get cloudy from all the people who live here. And the rain washed fertilizers from the farms into the water. The water became green with algae. When algae die, they fall to the bottom and become mud. Now the mud is pumped out of the water, and water fleas are used to eat the algae. The water is getting clear again. But there's another big problem. When the water quality is bad, water plants die and no longer protect the river banks. They're washed away – about three metres in ten years. Now the Broads Authority has covered the banks with asphalt and planted reeds to protect the banks. Then there are the farmers. They once kept cows and horses on the marshland. But in the 1970s the farmers began to drain the marshland. They built roads and barns and grew crops. Now the Broads Authority has stopped this. It pays farmers money if they keep their marshland for their cows and horses.

MARY But aren't tourists the biggest problem?

GUIDE Right. Almost two million tourists visit the Broads every year. There are many activities for them, like sailing, angling, walking and watching birds and insects. With so many visitors, car parking and litter are big problems. But I think we should start our tour now. I'd like to show you some beautiful birds and insects. If we're lucky, we'll see a swallowtail, Britain's largest butterfly.

3 The butterfly project

Use a library or the Internet to find out about the life cycle of the swallowtail butterfly. Collect pictures of the different stages from the caterpillar to the butterfly. Now write a short caption for each of the photos like the one below.

Swallowtails are the largest butterflies in the world. The butterfly's cycle from egg to adult butterfly is about 30 days.

4 Tourism in the Broads

a Make a list of the four big problems for the Norfolk Broads. Which is the biggest? Why?
b What kinds of jobs can people find in the tourist industry? Make a network.
c What tourist activity would you like to do most? Explain your answer.

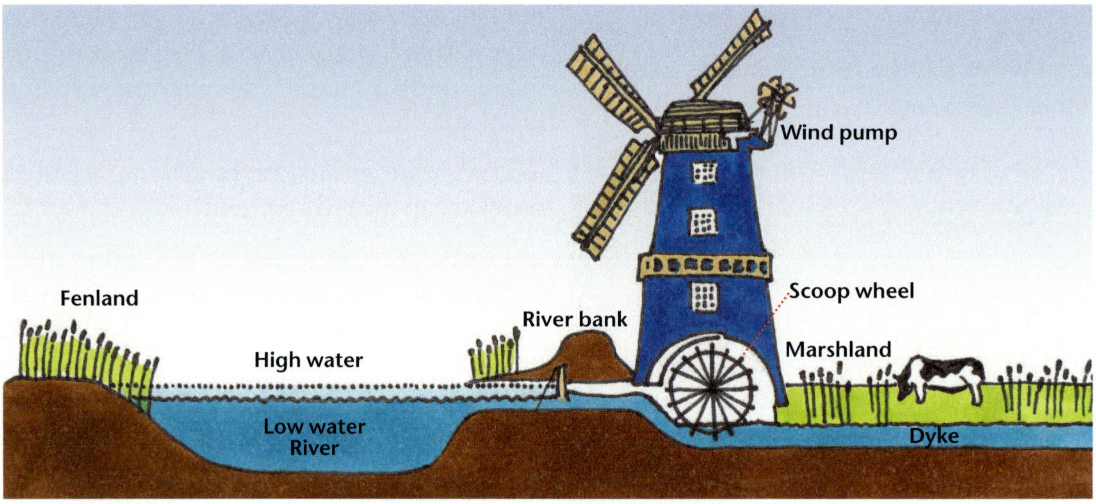

5 The marshes

Look at the diagram above and describe it. If you put these sentences (a–h) in the correct order, they can help you.

a However, ground water and rain water must be pumped into the river.
b When there's high water in the river, the marsh is blocked off, so it's safe.
c The grazing marshland in the Norfolk Broads is usually lower than the river.
d To protect the marshes from flooding by the sea, river banks are built first.
e The pumping was done by wind pumps first. Today electric pumps are used.
f Cows and horses can graze on the marshland without problems.
g The water flows to the sea at Great Yarmouth.
h Large scoop wheels lift the dyke water into the river.

6 Threats to the Broads

a Look at this diagram carefully and describe it.

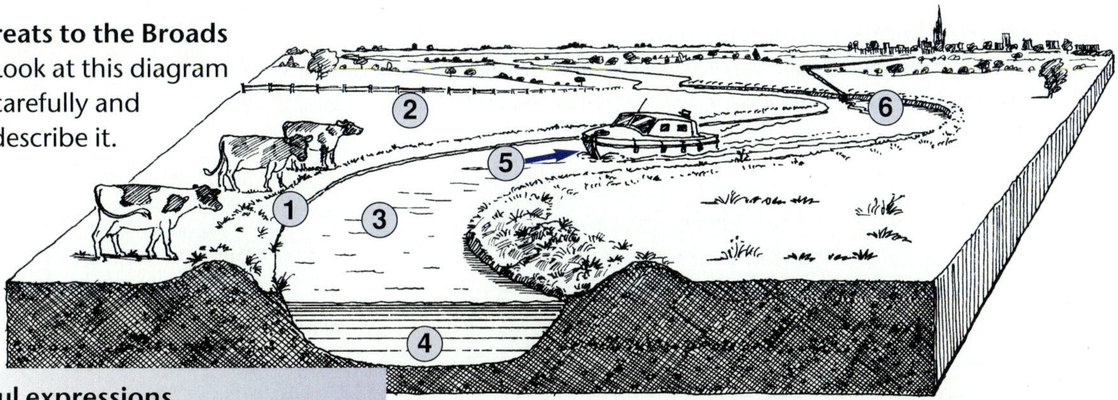

Helpful expressions

- The diagram shows part of the countryside of … It shows that the … is flat/hilly/…
- I can see a river / a motor cruiser / …
- There are a lot of / nearly no / only a few / … bushes/…
- The cows in the diagram show that …

b Now match the numbers in the diagram above (1–6) to these sentences:

a The banks are washed away by waves from motor boats.
b The water gets cloudy.
c Fertilizers from fields pollute the water.
d Sewage from towns gets into the water.
e Algae grow in the water, die and form mud.
f Cows walk on the banks and make them weak.

7 The future of the Broads
Look at the eight people below. They're all interested in the future of the Norfolk Broads.

a Write one sentence about what each person might think about the future of the Broads.
b Find pairs of people who might have similar views about the future of the Broads. Explain your answer.
c Find pairs of people who might disagree about the future of the Broads. Explain your answer.

8 The Hill Family's summer holidays

Mrs Hill and her family want to spend their holidays on a motor cruiser in the Norfolk Broads. They've studied a lot of brochures to find the right boat. Finally they've decided to hire this cruiser for two weeks.

a Describe how the cruiser is equipped.
b Imagine you and your family are on holiday on this cruiser. Where does everyone sleep?
c What do you do during the day on the cruiser?
d What food do you take with you for a one-week trip?
e Use the map of the Norfolk Broads on page 36 and plan a route.

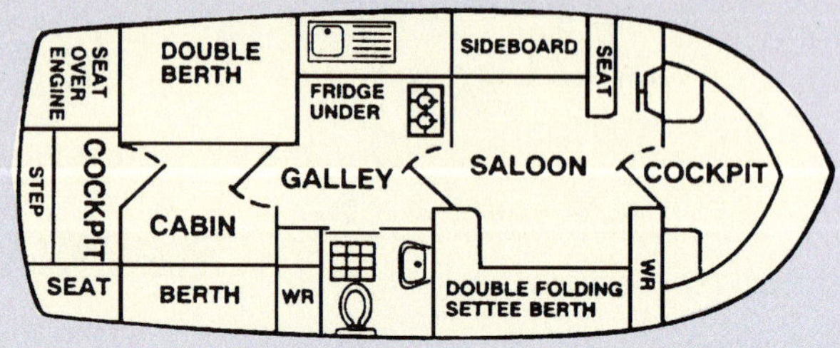

The cruiser is comfortable for up to four people and one child. The saloon and driving position have excellent visibility and a sliding roof for fresh air and sunshine. Your boat has both colour TV and heating.

Length: 30 ft *Beam:* 10 ft 6 in *Headroom:* 6 ft 2 in

9 The Broadland Code

When the Hills went to get their cruiser, the owner gave them a copy of the Broadland Code.

Study the Broadland Code and explain the rules and the reasons for them in your own words.

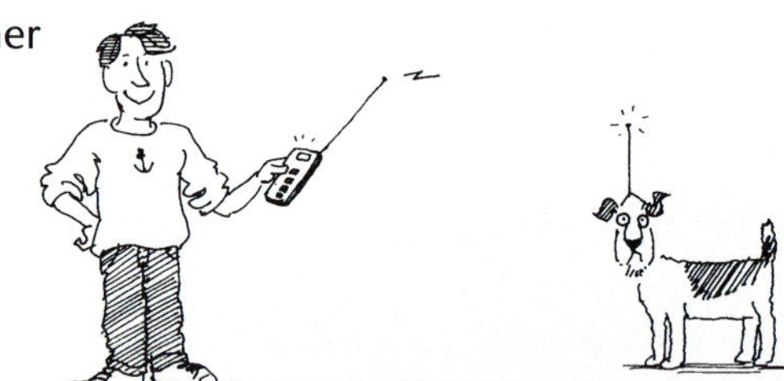

CONTROL YOUR DOG.

RESPECT OTHER BOAT USERS.

DON'T DISTURB WILDLIFE.

DON'T DROP LITTER.

DON'T SPILL OIL.

Talking about …

When you want to talk about maps, pictures, tables etc., the following expressions can be helpful.

… maps

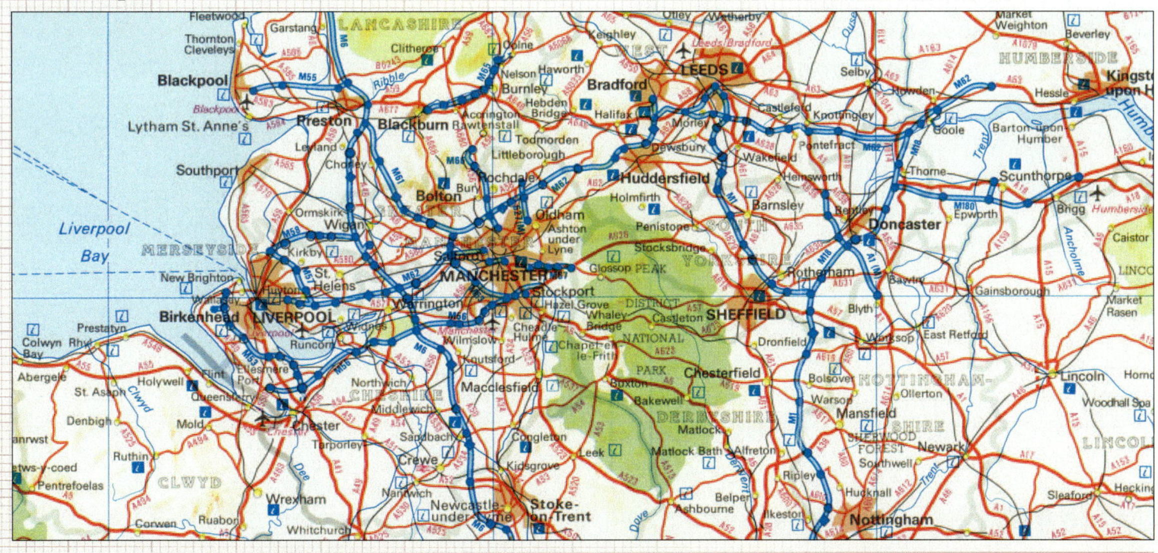

- This is a map of … / a map that shows a part of …
- It's a street/road/physical/political map.
- It's a modern/historical map. / It's a map that shows the situation during …
- … is in the north/east/south/west/north-east/south-east/south-west/ north-west of …
- … is / is located / is situated …
- … is about … km away from …
- … is near … / next to … / opposite … / between … and …
- … is on the road to … / on the coast / on the river.
- The A… runs between … and …

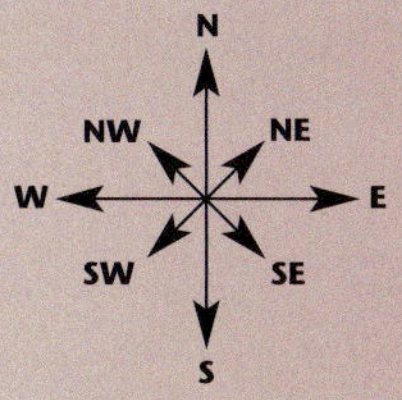

… pictures

- This is a drawing/photo/diagram/sketch of …
- The picture shows / illustrates / tells us about … It shows that …
- I can see … at the top/bottom on the left/right.
- In the foreground/background / at the front/back, there's / there are …
- There's / There are … in the middle / at the (very) edge of … / between … and …
- The people/… in the picture/… could be …
- The … probably is/are/was/were … Perhaps …
- The … look/s like … They/It could be made of …
- When you look at the …, you can guess that …
- The picture tells you something about …, because …

Talking about …

… tables

POPULATION OF CITIES IN ENGLAND, UK, 1981–2002 (Excerpt)			
Place	**Population** (in 1,000)		
	1981	**1991**	**2002**
London	6,569.3	6,638.1	7,393.8
Liverpool	538.8	481.8	466.6
Manchester	437.6	402.9	394.3
Chester	80.2	80.1	80.8

- This table shows the population/… of some cities in …
- The smallest / biggest / second biggest … is …
- There are more/fewer people in … than in …
- In … there are twice / three times / four times as many people as in …
- The number of people has gone up/down steadily/quickly/slowly / has remained (almost) level from 1981 to 2002 / …
- … has/had a population of … in 1981/…
- In 1981/… 80,200/… people lived in … Ten/… years later the number of people was/is …

… charts

Coffee exports from major countries (1999)

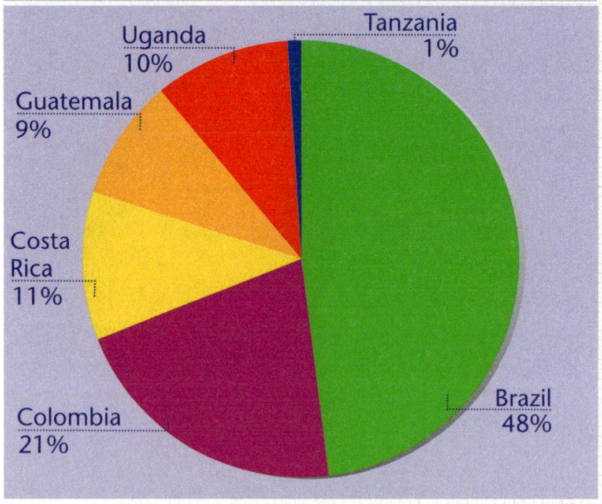

- This is a pie chart about …
- It shows six countries that export …
- … is the (second/third/fourth/…) biggest coffee exporter.
- … exports most / least / a lot of / (only) … % of all coffee.
- … exports more than twice / three times / four times as much coffee as …
- … exports nearly as much coffee as …
- Together, … export as much coffee as …
- Most countries are in … Only Uganda and Tanzania are … countries.

… climate graphs

Combination rainfall and temperature graphs for London

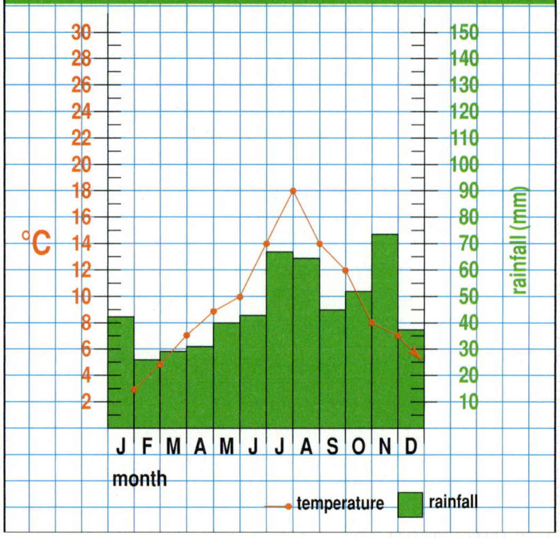

- This is a climate graph for …
- The graph shows the rainfall/precipitation and the temperature of …
- The warmest/coldest month of the year is …
- The highest/greatest/lowest/maximum/minimum rainfall/temperature is in …
- Rainfall/Temperatures in … is/are below/under/above/over/around … °C / mm.
- Temperatures fall/rise steadily/quickly/slowly in … / between … and …
- There isn't much / There's a big difference between … and …
- The total amount of rainfall per year is … mm.
- The mean annual temperature is … °C.

… dates

HISTORY OF THE OLYMPIC GAMES

776 BC	*Beginning of Ancient Olympics*
5th and 4th centuries BC	*Golden Age of Olympics*
393 AD	*Emperor Theodosius I ends Olympics*
6th April, 1896	*Pierre de Coubertin starts Modern Olympics*
1912	*First sportswomen allowed to take part*
1924	*Beginning of Winter Olympics*
1939–1945	*World War II; no Games in 1940 and 1944*
from 1992	*Separation of Winter and Summer Olympics*
1994	*First separated Winter Olympics*

- In (the year) seven seventy-six/…
- In/During the fifth and fourth/… centuries BC …
- In three nine three AD / three hundred and ninety-three AD …
- On the sixth of April eighteen (hundred and) ninety-six …
- From nineteen thirty-nine to nineteen forty-five …
- Since nineteen ninety-two …
- The first time when/that … was in …
- For … years/decades/centuries …
- About/Around … years ago …
- About/Around … years after/before 1896 / …
- … continued/lasted for … years / from … until …
- From that time on …
- In the following … years …

… texts

Herculaneum. So he put the warships to sea to help them.
Everyone was leaving the danger zone; my uncle showed no fear and sailed into it. All the time, he watched each new phase of the eruption; a slave wrote down all the details for him. Now ashes were falling around them, more and hotter when they came nearer the shore, together with stones. Suddenly they could not go on because the shore was full of stones and ashes. The people on the ship wanted to go back to Misenum, but my uncle, the admiral, told them to turn south to Stabiae, to save his friend Pomponianus. He landed there and took his frightened friend in his arms.

- The text is about / tells the story of …
- First …; later …; then …; next …; finally …
- First the text tells you about …, then there's a paragraph on … and in the end it says that …
- Lines 15 to 17 /… describe / explain / give you information on / …
- You can see that the uncle is … because in lines 21 to 23 he tells the people to …

- The heading/subheading mentions …
- It's in **bold** / *italic* / normal print / CAPITAL LETTERS.
- When you compare the text with the picture/map, …

- My first/… section goes from line … to line …
- I don't understand the word "shore" in line 17. / What does "shore" in line 17 mean?
- How do you pronounce the name in line 23?

Chapter 7

Mutiny on the Bounty

Many books have been written and many films have been made about the mutiny on the Bounty. Have you read or seen one? Tell the class what you know about this true story from the days of the great sailing ships before you read the text below.

The Bounty's captain, William Bligh, is showing a new officer his ship. This is what he tells him:

"She's a very fine ship, three masts, you see. We've got a crew of 45 men. Come down below deck. This is my cabin. It isn't very big, I'm afraid. The problem is, I've had to make room for the cargo. Yes, most of my cabin will soon be a little garden! We're sailing to Tahiti, you know – lovely island, Tahiti – and we must collect lots of breadfruit plants there. Well, those plants will live in my cabin while we sail to Jamaica! They need the breadfruit in Jamaica to feed the slaves, they say. Anyway, let's go on.

This is the room where the men sleep, and when you look past these stairs, you can see into the galley. Ah, the cook is busy with the evening meal. Mmm, smells good, doesn't it? Now, down these stairs on the right is the main hold. We have barrels and barrels of meat and water there. We keep the ship's biscuit at the other end of the ship in the breadroom. It's good and dry there. Now, on the left here is the ship's carpenter's cabin. He always has something to repair when we meet the first storms.

Well, sir, let's return to the deck. Ah, yes, look at the ship's launch here – a lovely little boat. You know, I always think, if there's any trouble with the men, I'll just put them in the launch and away they go, out to sea. Ha! Ha!"

1 Make a list of the parts of the ship which the captain showed the officer. Then match the names to the numbers on the picture.

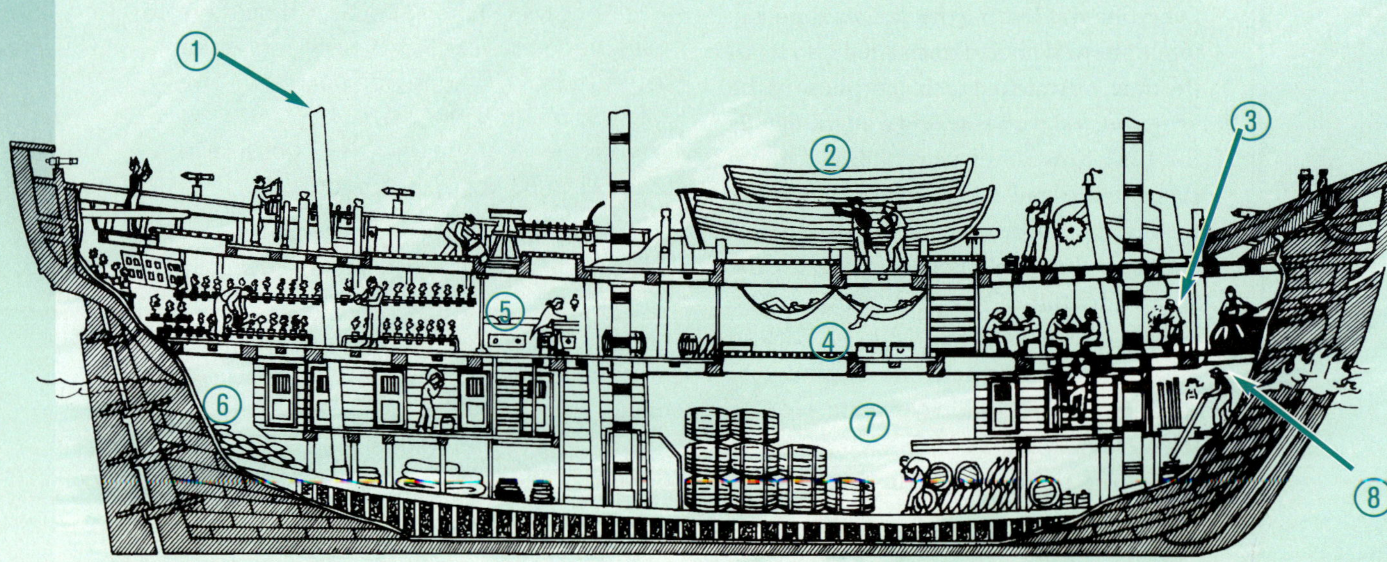

The voyage of the Bounty

Captain Bligh made many voyages all over the world in the sailing ship Bounty. The most famous of these voyages is the one that started from Portsmouth, England in 1787. The map below shows which countries the ship visited on the voyage.

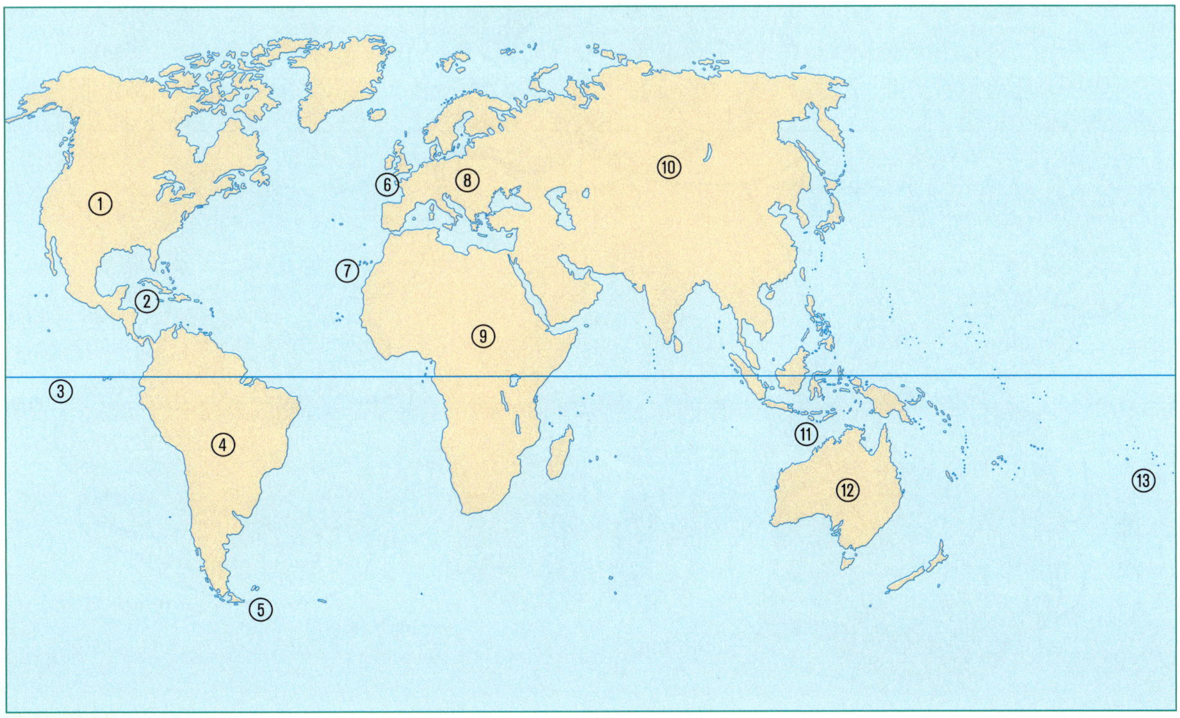

2 Match the following places to the numbers on the map. You can also use the map to follow the Bounty's voyage (see task 3).

Africa	Cape Horn	Portsmouth
Asia	Jamaica	
Australia	Tahiti	the equator
Europe	Tenerife	
North America	Timor	
South America		

3 Study the diary on the right. Then write the text below in your exercise book and fill in the correct numbers.

The voyage of the Bounty took place about (1) … years ago. In all, the ship was at sea for (2) … months before it reached Tahiti. It took (3) … weeks just to reach Cape Horn. The crew spent (4) … days in the storms at the Cape. They finally reached Tahiti after (5) … more months at sea. In the following (6) … months, they lived happily on Tahiti. The crew loved the warm sun, the fresh food, the friendly people – some sailors even found wives on the island! There was trouble when the captain gave the order: "Back on ship!" (7) … days later came the mutiny.

23rd Dec., 1787	*Set sail from Portsmouth, England*
10th Jan., 1788	*Visited Tenerife*
7th Feb.	*Crossed the equator*
23rd Mar.	*Reached Cape Horn – storms!*
22nd Apr.	*Had to turn back & go east – storms too bad*
26th Oct.	*Landed on Tahiti*
4th Apr., 1789	*Left Tahiti for Jamaica*
28th Apr.	*MUTINY!*

4 Collect all the phrases from task 3 to talk about (past) time. Use them to make more sentences about the voyage.

The reasons for the mutiny

5 What made the crew so unhappy that they were ready to mutiny? Remember the voyage of the Bounty and look at the pictures of the sailors' lives. Then write down all the reasons that you can think of:

- They had to sleep in terrible …
- They weren't allowed to …
- The captain often …
- …

The mutiny started when Captain Bligh counted his coconuts and found that some were missing. He shouted at everyone, even at his friend, Fletcher Christian, who then decided to leave the ship. But the sailors had had enough. "Don't leave us, Christian. Help us," they said. "Be our leader in the mutiny."

The mutiny

6 Match the captions (a–f) to the pictures (1–6).

a The men in the launch beg the mutineers for food, drink and navigational instruments.

b With his hands behind his back, they take him up on deck.

c Some of the mutineers want to fire on the launch. Captain Bligh's launch leaves the mother ship.

d Captain Bligh and 18 of his men climb into the launch.

e 'What do we do now?' the mutineers argue.

f The mutineers go into Captain Bligh's cabin while he's sleeping.

7 Look at the pictures again. What do you think the men are saying?

1	Captain Bligh:	Hello. Is it morning already? Who … What's this, in God's name?
	Christian:	It's a mutiny, Captain – a mutiny!
	Captain Bligh:	…
2	…	

After the mutiny

Helpful expressions

- g = gram(s)
- ml = millilitre(s)
- sq km = square kilometre(s)

▶ p. 55

8 Use these numerical expressions to complete the text below:
57 g • 7 m • 48 km • 14th • 142 ml • 6,280 km • 2 m
33,912 sq km

The launch was only (1) … long and (2) … wide. Bligh and his 18 men were alone in the open boat. First, Bligh decided to sail to the island of Tofua, (3) … away, to get more food and drink. But the islanders attacked the boat and killed one of the men. After that, the crew stayed at sea and visited no more islands.
On the voyage, each man was allowed to eat only (4) … of bread per day and have (5) … of water to drink. Day after day, they saw only the sea. Could they survive? Captain Bligh used navigational instruments very cleverly. He now wanted to reach Timor, an island of (6) …, to the north of Australia. From there they could get home. After 43 days, on (7) … June, the captain saw that they were safe. There before them lay Timor, behind them a journey of (8) … at sea!

9 Now practise reading the text out loud. Make sure you say the numbers correctly.

Topic

Costume Quiz

Look at these people in historical costumes. The boxes below tell you who they are (1–7) and when they wore these clothes (a–g). Choose one caption from each box for each costume (I–VII).

1 Viking soldiers	a in the New Stone Age
2 a German farmer	b in the 8th and 9th century
3 an English gentleman's family	c about 3,500 years ago
4 a rich Egyptian lady	d from before 1789
5 American pioneers	e about 400 years ago
6 a French baron	f from the 1890s
7 a European hunter	g until the 14th century

I

II

III

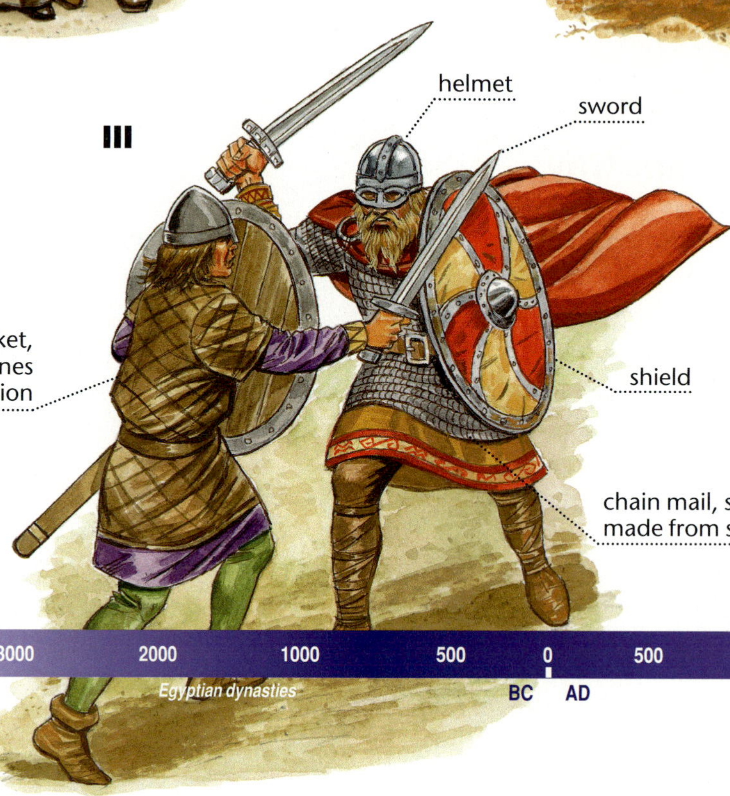

5000 4000 3000 2000 1000 500 0 500 1000 1100
New Stone Age — Egyptian dynasties — BC AD — Vikings conquer Europe

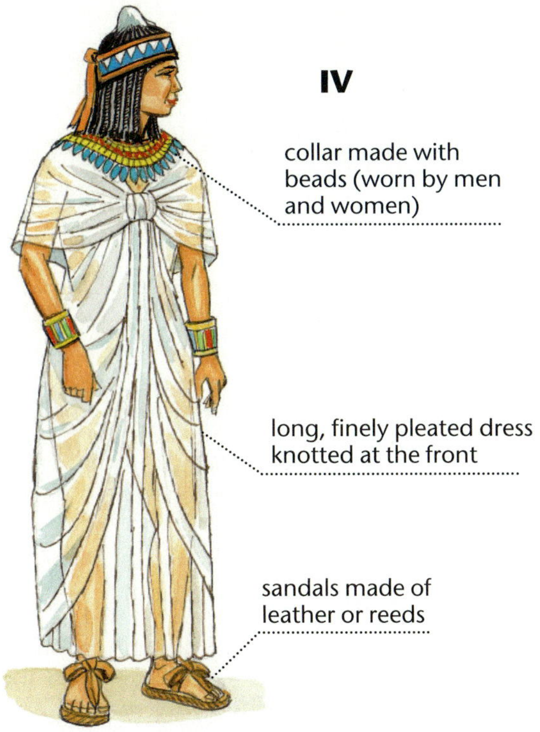

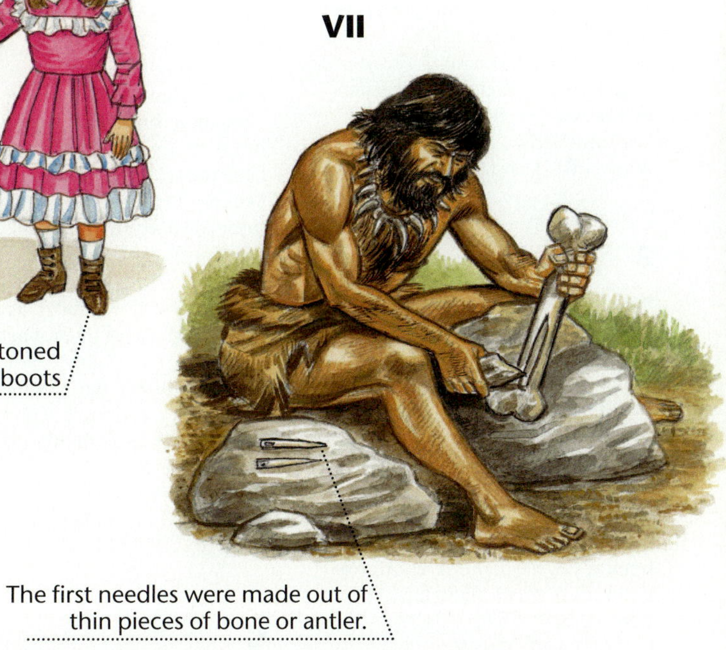

Helpful expressions

- in/from 4000 BC
- in the first century BC
- in the year 800 AD
- in the 1620s
- in 1789

1200 1300 1400 1500 1600 1700 1800 1900 2000

Middle Ages *1620 Pilgrims settle in America* *1789 French Revolution*

Vocabulary

A

above [ə'bʌv] über; oben; obige(r, s)
acre ['eɪkə] *Flächenmaß (0,405 ha)*
active ['æktɪv] aktiv
AD (Anno Domini) nach Christus
adult ['ædʌlt] ausgewachsen
adventurer [əd'ventʃərə] Abenteurer/in
afraid [ə'freɪd]: **be afraid** fürchten
age Zeitalter
algae ['ældʒiː] Algen
alive [ə'laɪv]: **stay alive** am Leben bleiben
all over again wieder von vorn
allergic [ə'lɜːdʒɪk] allergisch
among [ə'mʌŋ] unter
amount [ə'maʊnt] Menge
amphitheatre ['æmfɪθɪətə] Amphitheater
ancient ['eɪnʃənt] antik
angel ['eɪndʒl] Engel
angle ['æŋgl] Winkel
angler ['æŋglə] Angler/in
angling ['æŋglɪŋ] Angeln
announcer [ə'naʊnsə] Ansager/in
annual ['ænjuəl] Jahres-
antler ['æntlə] Geweihstange
anyway ['eniweɪ] wie auch immer
April Fool's Day der 1. April
apron ['eɪprən] Schürze
arable ['ærəbl] Ackerland
argue ['ɑːgjuː] sich streiten
armpit ['ɑːmpɪt] Achselhöhle
arrow ['ærəʊ] Pfeil
artificial [ˌɑːtɪ'fɪʃl] künstlich
as als, während
ash [æʃ] Asche
asleep [ə'sliːp]: **be asleep** schlafen
asphalt ['æsfælt] Asphalt
atmosphere ['ætməsfɪə] Atmosphäre
attack [ə'tæk] befallen; angreifen
attractive [ə'træktɪv] attraktiv, ansprechend
automatic [ˌɔːtə'mætɪk] automatisch

B

back Rückseite; **in the back** hinten
background Hintergrund
backpack Rucksack
bacteria [bæk'tɪəriə] Bakterien
bad luck Pech
baggy ['bægi] weit *(geschnitten)*
bait [beɪt] Köder
baking powder ['beɪkɪŋ paʊdə] Backpulver
baking sheet ['beɪkɪŋ ʃiːt] Backblech
baking tin ['beɪkɪŋ tɪn] Backform
balloon [bə'luːn] (Luft-)Ballon
bank Ufer
barley ['bɑːli] Gerste
barn [bɑːn] Scheune
barometer [bə'rɒmɪtə] Barometer
barracks ['bærəks] Kaserne
barrel ['bærəl] Fass
basis ['beɪsɪs]: **on a daily basis** tageweise
bat [bæt] Fledermaus
batter ['bætə] Teig
bay [beɪ] Bucht
BC (before Christ) vor Christus
bead [biːd] Perle
beam [biːm] Breite *(beim Schiff)*
bean [biːn] Bohne
became [bɪ'keɪm] *si. past „become"*
become [bɪ'kʌm] *(became, become)* werden
bedside light [ˌbedsaɪd 'laɪt] Nachttischlampe
beer [bɪə] Bier
beg [beg] betteln, anflehen
bell [bel] Glocke
below unter; unten
berth [bɜːθ] Koje; Kajütenbett
biology [baɪ'ɒlədʒi] Biologie
biscuit: **ship's biscuit** Schiffszwieback
bit *si. past/Part. Perf. „bite"*
bite [baɪt] *(bit, bit)* beißen; anbeißen
blanket ['blæŋkɪt] Decke
blend [blend] vermischen
block [blɒk]: **block off** blockieren
blood [blʌd] Blut
blow one's top *(blew, blown)* in die Luft gehen
boatyard ['bəʊtjɑːd] Bootshandlung
body ['bɒdi] Leiche
boil [bɔɪl] **1.** kochen; **2.** Furunkel
boot [buːt] Stiefel
born [bɔːn]: **be born** geboren sein/werden
bottom: **at the bottom left/right** unten links/rechts
bow [bəʊ] Bogen
brave [breɪv] tapfer
break [breɪk] *(broke, broken)* (auf)-brechen; **break up** zerkleinern
breathe [briːð] atmen
breeches ['brɪtʃɪz] Kniehose
broken ['brəʊkən] *Part. Perf. „break"*
bubonic plague [bjuːˌbɒnɪk 'pleɪg] Beulenpest
bucket ['bʌkɪt] Eimer
buckle ['bʌkl] Schnalle, Spange
build (up) [bɪld] *(built, built)* (auf)-bauen
builder ['bɪldə] Bauarbeiter/in
built [bɪlt] *si. past/Part. Perf. „build"*
bury ['beri] begraben
bush [bʊʃ] Busch
butterfly ['bʌtəflaɪ] Schmetterling
button ['bʌtn] knöpfen
by mal

C

cabin ['kæbɪn] Kabine
Canadian [kə'neɪdiən] Kanadier/in
cannibal ['kænɪbl] Kannibale, Kannibalin
cap [kæp] Haube
cape [keɪp] Kap
capital letter Großbuchstabe
car parking Parken
cargo ['kɑːgəʊ] Fracht, Ladung
carpenter ['kɑːpəntə] Zimmermann
catch [kætʃ] **1.** *(caught, caught)* fangen; sich zuziehen; **2.** Fang
caterpillar ['kætəpɪlə] Raupe
Catholic ['kæθlɪk] Katholik/in
caught [kɔːt] *si. past/Part. Perf. „catch"*
cauldron ['kɔːldrən] Kessel
cave [keɪv] Höhle
cemetery ['semətri] Friedhof
centimetre ['sentɪmiːtə] Zentimeter
certain ['sɜːtn] bestimmt
chain mail ['tʃeɪn meɪl] Kettenhemd, -panzer
chalk [tʃɔːk] Kreide
chart [tʃɑːt] Tabelle, Diagramm
china ['tʃaɪnə] Porzellan
cinnamon ['sɪnəmən] Zimt
climate ['klaɪmət] Klima
close dicht, nahe
cloth [klɒθ] Stoff, Tuch
cloudy ['klaʊdi] bewölkt, wolkig
coal [kəʊl] Kohle
coconut ['kəʊkənʌt] Kokosnuss
code [kəʊd] Regeln, Kodex
cold [kəʊld] Erkältung; **be cold** frieren
collapse [kə'læps] zusammenbrechen
collar ['kɒlə] Kragen
colouring ['kʌlərɪŋ] Farbstoff
combine [kəm'baɪn] kombinieren
comfortable ['kʌmftəbl] angenehm, bequem
commercial [kə'mɜːʃl] Werbespot
company ['kʌmpəni] Firma, Gesellschaft
compass ['kʌmpəs] Kompass
condense [kən'dens] kondensieren
conditions [kən'dɪʃnz] Bedingungen
confession [kən'feʃn] Beichte
connection [kə'nekʃn] Beziehung, Verbindung
conquer ['kɒŋkə] erobern
construction [kən'strʌkʃn] Konstruktion
contact ['kɒntækt] Kontakt
continent ['kɒntɪnənt] Kontinent
continue [kən'tɪnjuː] andauern
control unter Kontrolle halten
cool (sich) abkühlen
copy Kopie
corner ['kɔːnə] Ecke
cotton ['kɒtn] Baumwolle
cough [kɒf] husten
countryside ['kʌntrisaɪd] Landschaft
courtyard ['kɔːtjɑːd] Hof
cover ['kʌvə] bedecken; **be covered in** bedeckt sein mit

crash-land ['kræʃlænd] eine Bruchlandung machen
crater ['kreɪtə] Krater
cream-coloured ['kri:mkʌləd] cremefarben
crew [kru:] Besatzung
crocodile ['krɒkədaɪl] Krokodil
crop [krɒp] Ernte; Feldfrucht
cross section Querschnitt
crown [kraʊn] krönen
cruiser ['kru:zə]: **motor cruiser** Motorboot, -jacht
culture ['kʌltʃə] Kultur
cup [kʌp] Tasse
curtain ['kɜ:tn] Vorhang, Gardine
cycle Zyklus

D

daily ['deɪli] täglich
dairy cow ['deəri kaʊ] Milchkuh
dark [dɑ:k] Dunkelheit
day after day Tag für Tag
daytime Tag
dead Tote(r)
deadly ['dedli] tödlich
dear: **Oh dear!** Oje!
death [deθ] Tod
decade ['dekeɪd] Jahrzehnt
deport [dɪ'pɔ:t] ausweisen
desert ['dezət] Wüste
design [dɪ'zaɪn] entwerfen
dessert [dɪ'zɜ:t] Nachtisch, -speise
devil ['devl] Teufel
diagram ['daɪəgræm] Diagramm, schematische Darstellung
diameter [daɪ'æmɪtə] Durchmesser
difficult ['dɪfɪkəlt] schwierig
dig [dɪg] *(dug, dug)* graben
direction [də'rekʃn] Richtung
dirt [dɜ:t] Schmutz, Dreck
disagree [ˌdɪsə'gri:] nicht übereinstimmen
disaster [dɪ'zɑ:stə] Katastrophe
disturb [dɪ'stɜ:b] stören
dome [dəʊm] Kuppel
double ['dʌbl] Doppel-
drain [dreɪn] entwässern
draw *(drew, drawn)* ziehen
drinking straw ['drɪnkɪŋ strɔ:] Trinkhalm, Strohhalm
driving position (Boots-)Führerstand
drought [draʊt] Dürre
dry trocknen
during ['djʊərɪŋ] während
dyke [daɪk] Deich, Damm
dynasty ['dɪnəsti] Dynastie

E

earth [ɜ:θ] Erde
economy [ɪ'kɒnəmi] Wirtschaft
ecosystem ['i:kəʊsɪstəm] Ökosystem
edge [edʒ]: **at the (very) edge** am (äußersten) Rand
effect [ɪ'fekt] Wirkung
Egyptian [i'dʒɪpʃn] Ägypter/in
elsewhere [ˌels'weə] woanders(hin)
emperor ['empərə] Kaiser
employ [ɪm'plɔɪ] beschäftigen
end enden; beenden; **in the end** schließlich
enough [ɪ'nʌf] genug
envelope ['envələʊp] (Brief-)Umschlag
equator [ɪ'kweɪtə] Äquator
equip [ɪ'kwɪp] ausrüsten, ausstatten
equipment [ɪ'kwɪpmənt] Ausrüstung, Ausstattung
erupt [ɪ'rʌpt] ausbrechen
eruption [ɪ'rʌpʃn] Ausbruch
etc. (et cetera) [et 'setərə] usw. (und so weiter)
European [ˌjʊərə'pi:ən] **1.** europäisch; **2.** Europäer/in
evaporate [ɪ'væpəreɪt] verdunsten
even ['i:vn] **1.** sogar; **2.** eben, glatt
event [ɪ'vent] Ereignis
excellent ['eksələnt] ausgezeichnet, hervorragend
excerpt ['eksɜ:pt] Auszug, Ausschnitt
excited [ɪk'saɪtɪd] aufgeregt
excercise Training
expect [ɪk'spekt] erwarten
export **1.** [ɪk'spɔ:t] exportieren; **2.** ['ekspɔ:t] Export
exporter [ɪk'spɔ:tə] Exporteur/in
extreme [ɪk'stri:m] extrem

F

fact [fækt] Tatsache, Fakt
fallout Niederschlag
far: **on the far left/right** ganz links/rechts
farmer ['fɑ:mə] Bauer/Bäuerin
fattening ['fætnɪŋ] dick machend
favourite Favorit/in
feather ['feðə] Feder
feed [fi:d] (Vieh-)Futter
fenland ['fenlænd] Marschland
ferment [fə'ment] zur Gärung bringen
fertile ['fɜ:taɪl] fruchtbar
fertilizer ['fɜ:təlaɪzə] Dünger, Düngemittel
fever ['fi:və] Fieber
figure ['fɪgə] Zahl
finally ['faɪnəli] endlich, schließlich
fire on sb./sth. auf jn./etw. schießen
first of all zuallererst
fishing line Angelschnur
fix [fɪks] befestigen
flame [fleɪm] Flamme
flare [fleə] Leuchtrakete
flask [flɑ:sk] (Reise-)Flasche
flat flach
flea [fli:] Floh
fleet [fli:t] Flotte
flip up [flɪp 'ʌp] hochschnellen
flog [flɒg] auspeitschen
flood [flʌd] **1.** überschwemmen; **2.** Überschwemmung
flour ['flaʊə] Mehl
flow [fləʊ] fließen
flower ['flaʊə] Blume
fog [fɒg] Nebel
fold [fəʊld] **1.** falten; **2.** sich zusammenklappen lassen
for nach; **for nothing** umsonst
foreground ['fɔ:graʊnd] Vordergrund
fork [fɔ:k] Gabel
form bilden
forum ['fɔ:rəm] Forum
fresh [freʃ] frisch
fridge [frɪdʒ] Kühlschrank
friendly ['frendli] freundlich
frightened ['fraɪtnd] verängstigt
frightening ['fraɪtnɪŋ] Furcht erregend
front Vorderseite
full-time Ganztags-

G

galley ['gæli] Schiffsküche, Kombüse
gate [geɪt] Tor
get *(got, got)*: **get infected** sich infizieren; **get lost** sich verlaufen; **get married** heiraten; **get out of bed** aufstehen; **get sth. ready** etwas fertig machen
giggle ['gɪgl] kichern
glove [glʌv] Handschuh
go *(went, gone)*: **go on** weiterfahren
God [gɒd] Gott
goods [gʊdz] Waren, Güter
government ['gʌvənmənt] Regierung
gram [græm] Gramm
graph [græf] grafische Darstellung
grass [grɑ:s] Gras
grave [greɪv] Grab
graze [greɪz] grasen
great [greɪt] groß
grew [gru:] *si. past „grow"*
grind [graɪnd] *(ground, ground)* mahlen
groan [grəʊn] stöhnen
groin [grɔɪn] Leistengegend
ground [graʊnd] Boden, Erde; Grund-
grow [grəʊ] *(grew, grown)* wachsen; anbauen
grown [grəʊn] *Part. Perf. „grow"*
guard [gɑ:d]: **guard tower** Wachturm

H

hall [hɔ:l] Halle
ham [hæm] Schinken
hammock ['hæmək] Hängematte
hang [hæŋ] *(hung, hung)* hängen
harvest ['hɑ:vɪst] **1.** ernten; **2.** Ernte
headline Schlagzeile
headroom *(lichte)* Höhe
headwind Gegenwind
healthy ['helθi] gesund
heat [hi:t] **1.** erhitzen; **2.** Hitze
heating ['hi:tɪŋ] Heizung

heaven ['hevn] Himmel
heavy ['hevi] schwer *(von Gewicht)*
hectare ['hekteə] Hektar
helpful ['helpfl] hilfsbereit
herb [hɜ:b] Kraut
heritage ['herɪtɪdʒ] Erbe
hide [haɪd] *(hid, hidden)* (sich) verstecken
hilly ['hɪli] hügelig
hire ['haɪə] mieten
historian [hɪ'stɔ:riən] Historiker/in
historic [hɪ'stɒrɪk] historisch *(bedeutend)*
historical [hɪ'stɒrɪkl] historisch, geschichtlich
hold Lade-, Frachtraum
home Haus, Heim; Heimat-
hook [hʊk] Haken
horse [hɔ:s] Pferd
horseback riding Pferdereiten
how to do sth. wie man etwas tun kann/soll
however [haʊ'evə] jedoch, aber
hung [hʌŋ] *si. past/Part. Perf. „hang"*
hunger ['hʌŋgə] Hunger
hunter ['hʌntə] Jäger/in
hutch [hʌtʃ] Hütte

I

ice [aɪs] Eis
if wenn
illness ['ɪlnəs] Krankheit
illustrate ['ɪləstreɪt] illustrieren
import [ɪm'pɔ:t] importieren, einführen
inch [ɪntʃ] *Längenmaß (2,54 cm)*
independent [,ɪndɪ'pendənt] unabhängig
industry ['ɪndəstri] Industrie
infect [ɪn'fekt] infizieren
insect ['ɪnsekt] Insekt
inside [,ɪn'saɪd] in … (hinein)
interested ['ɪntrəstɪd]: **be interested (in)** sich interessieren (für)
invent [ɪn'vent] erfinden
iron ['aɪən] Eisen
islander ['aɪləndə] Inselbewohner/in
italic [ɪ'tælɪk] kursiv
ivory ['aɪvəri] Elfenbein

J

jack-o-lantern (Kürbis-)Laterne
jam [dʒæm] Marmelade
Jewish ['dʒu:ɪʃ] jüdisch
journey ['dʒɜ:ni] Reise, Fahrt
jungle ['dʒʌŋgl] Dschungel
just like genau wie

K

key [ki:] Zeichenerklärung
kind Art, Sorte
knot [nɒt] knoten

L

landscape ['lændskeɪp] Landschaft
last **1.** dauern; **2.** das letzte Mal, zuletzt
launch [lɔ:ntʃ] Barkasse, Beiboot
lava ['lɑ:və] Lava
lay [leɪ] **1.** *(laid, laid)* legen; **2.** *si. past „lie"*
layer ['leɪə] Schicht
leader ['li:də] Anführer/in
leaf [li:f], *pl* **leaves** [li:vz] Blatt
least [li:st] am wenigsten
leather ['leðə] Leder
leave [li:v] *(left, left)* lassen; **leave for** aufbrechen nach
left [left] *si. past/Part. Perf. „leave"*
lemon grass ['lemənɡrɑ:s] Zitronengras
length [leŋθ] Länge
level ['levl]: **remain level** gleich bleiben
lie [laɪ] *(lay, lain)* liegen
lift [lɪft] (an)heben
lighter ['laɪtə] Feuerzeug
line: **line graph** Kurvendiagramm; **fishing line** Angelschnur
linen ['lɪnɪn] Leinen
litter ['lɪtə] Abfall
local ['ləʊkl] örtlich; einheimisch
located [ləʊ'keɪtɪd]: **be located** liegen; gelegen sein
lock Schleuse
long lange
look: **look after sb.** sich um jn. kümmern; **look up** nachschlagen
love: **in love** verliebt
low [ləʊ] niedrig, tief
lowland Tiefland
luck [lʌk]: **bad luck** Pech
lump [lʌmp] Klumpen
lung [lʌŋ] Lunge

M

made of aus (… gemacht)
maggot ['mægət] Made
magma ['mægmə] Magma
main [meɪn] Haupt-
maize [meɪz] Mais
major ['meɪdʒə] größere(r, s)
make *(made, made)*: **make room** Platz machen; **make sb. do sth.** jn. dazu bringen/zwingen, etwas zu tun; **make sure** sorgen für; kontrollieren
mark [mɑ:k] markieren
market ['mɑ:kɪt] Markt
marry ['mæri] heiraten
marsh [mɑ:ʃ] Sumpf
marshy sumpfig
mask [mɑ:sk] Maske
mast [mɑ:st] Mast
match: **with a pattern/… to match sth.** mit einem (genau) zu etwas passenden Muster/…
mature [mə'tʃʊə] ausgewachsen
maximum ['mæksɪməm] Höchst-; maximal

mean durchschnittlich
measure ['meʒə] messen
medal ['medl] Medaille
meet sth. *(met, met)* auf etwas treffen
member ['membə] Mitglied
mention ['menʃn] erwähnen
merchant ['mɜ:tʃənt] Kaufmann
Middle Ages Mittelalter
might [maɪt] könnte(n) vielleicht
milk melken
mill [mɪl] Mühle
millilitre ['mɪlili:tə] Milliliter
millimetre ['mɪlimi:tə] Millimeter
minimum ['mɪnɪməm] Mindest-; minimal
minus ['maɪnəs] minus
mission ['mɪʃn] Mission, Aufgabe
mix vermischen, verrühren
mixing bowl ['mɪxɪŋ bəʊl] Rührschüssel
mixture ['mɪkstʃə] Mixtur, Gemisch
moderate ['mɒdərət] mäßig
modern ['mɒdn] modern
moisture ['mɔɪstʃə] Feuchtigkeit
monument ['mɒnjʊmənt] Denkmal
motor cruiser ['məʊtə kru:zə] Motorboot, -jacht
movement ['mu:vmənt] Bewegung
mud [mʌd] Schlamm
mutineer [,mju:tə'nɪə] Meuterer/Meuterin
mutiny ['mju:tɪni] **1.** meutern; **2.** Meuterei

N

nasty ['nɑ:sti] böse
nature reserve ['neɪtʃə rɪ,zɜ:v] Naturschutzgebiet
navigational instrument [,nævɪ'geɪʃnəl 'ɪnstrʊmənt] Navigationsinstrument
neck [nek] Ausschnitt
needle ['ni:dl] Nadel
New Stone Age Jungsteinzeit
news [nju:z] Nachrichten
next [nekst] als Nächstes
nonsense ['nɒnsns] Unsinn
normal ['nɔ:ml] normal
note [nəʊt] Anmerkung
nothing: **for nothing** umsonst
numerical [nju:'merɪkl] Zahlen-
nut [nʌt] Nuss
nutmeg ['nʌtmeg] Muskatnuss

O

oats [əʊts] Hafer
ocean ['əʊʃn] Meer; Ozean
offer ['ɒfə] anbieten
olive ['ɒlɪv] Olive
Olympics [ə'lɪmpɪks] Olympiade
on: **from that time on** von jener Zeit an
once [wʌns] einmal
one-week einwöchige(r, s)

ooze [u:z] absondern
order Befehl
organic [ɔ:'gænɪk] organisch, naturgemäß
outer äußere(r, s), Außen-
outside [ˌaʊt'saɪd] außerhalb
oven ['ʌvn] Ofen, Backofen
over hinüber
owner ['əʊnə] Besitzer/in

P

packaging ['pækɪdʒɪŋ] Verpackung
packhorse ['pækhɔ:s] Packpferd
padded ['pædɪd] wattiert, gepolstert
pain [peɪn] Schmerz
paper ['peɪpə] Papier
paradise ['pærədaɪs] Paradies
parasol ['pærəsɒl] Sonnenschirm
part-time Halbtags-
passive ['pæsɪv] Passiv
past an ... vorbei
path [pɑ:θ] Pfad
pattern ['pætn] Muster
pen-friend Brieffreund/in
pepper ['pepə] Pfeffer
per [pə] pro; **per cent** [ˌpə 'sent] Prozent
perfect ['pɜ:fɪkt] perfekt
pesticide ['pestɪsaɪd] Pestizid
petticoat ['petɪkəʊt] Unterrock
physical ['fɪzɪkl] dinglich, physisch
pianist ['pɪənɪst] Pianist/in
pick [pɪk] aussuchen; pflücken
pie chart ['paɪ tʃɑ:t] Kreisdiagramm
pilgrim ['pɪlgrɪm] Pilger/in
pioneer [ˌpaɪə'nɪə] Pionier/in
pipe [paɪp] Rohr
piranha [pɪ'rɑ:nə] Piranha
pistol ['pɪstl] Pistole
plague [pleɪg] Seuche, Pest
plant [plɑ:nt] **1.** (an)pflanzen; **2.** Pflanze
Plasticine ['plæstəsi:n] Plastilin *(Knetmasse)*
pleated ['pli:tɪd] in Falten gelegt
plough [plaʊ] pflügen
pneumonic plague [nju:ˌmɒnɪk 'pleɪg] Lungenpest
pole [pəʊl] Pol
political [pə'lɪtɪkl] politisch
politics ['pɒlətɪks] Politik
pope [pəʊp] Papst
popular ['pɒpjələ] beliebt
population [ˌpɒpjʊ'leɪʃn] Bevölkerung(szahl)
possible ['pɒsəbl] möglich
potholing ['pɒthəʊlɪŋ] Höhlen erforschen
pour [pɔ:] (ein)gießen
pray [preɪ] beten
precipitation [prɪˌsɪpɪ'teɪʃn] Niederschlag
prepare [prɪ'peə] (sich) vorbereiten
pressure ['preʃə] Druck
prevent [prɪ'vent] verhindern
priest [pri:st] Priester/in
prison ['prɪzn] Gefängnis
private ['praɪvət] privat, Privat-
probably ['prɒbəbli] wahrscheinlich
produce [prə'dju:s] produzieren, herstellen
product ['prɒdʌkt] Produkt, Erzeugnis
properly ['prɒpəli] richtig
protect [prə'tekt] schützen
protection [prə'tekʃn] Schutz
public ['pʌblɪk] öffentlich
pulp [pʌlp] zerdrücken
pump [pʌmp] **1.** pumpen; **2.** Pumpe
punish ['pʌnɪʃ] bestrafen
punishment ['pʌnɪʃmənt] Strafe
pus [pʌs] Eiter
put to sea in See stechen, auslaufen (mit)

Q

quality ['kwɒləti] Qualität

R

railway ['reɪlweɪ] Eisenbahn
rainfall ['reɪnfɔ:l] (Regen-)Schauer; Niederschlag
rainy ['reɪni] regnerisch
rat [ræt] Ratte
raw [rɔ:] roh
reaction [ri:'ækʃn] Reaktion
recipe ['resəpi] (Koch-, Back-)Rezept
reconstruct [ˌri:kən'strʌkt] wieder aufbauen
record Rekord
reed [ri:d] Schilf, Ried
refrigeration [rɪˌfrɪdʒə'reɪʃn] Kühlung
region ['ri:dʒən] Region, Gebiet
relative ['relətɪv] Verwandte
release sb. (from) [rɪ'li:s] jn. entlassen (aus)
remain [rɪ'meɪn]: **remain level** gleich bleiben
remains Reste
remind sb. (of sth.) [rɪ'maɪnd] jn. (an etwas) erinnern
repair [rɪ'peə] reparieren
repellent [rɪ'pelənt] Schutzmittel
resolution [ˌrezə'lu:ʃn] Vorsatz
respect [rɪ'spekt] respektieren
result [rɪ'zʌlt] Ergebnis
retiree [rɪˌtaɪə'ri:] Rentner/in
return [rɪ'tɜ:n] zurückkehren
ribbon ['rɪbən] Band, Schleife
rich [rɪtʃ] reich
right: **right angle** rechter Winkel; **right away** sofort
ripe [raɪp] reif
rise [raɪz] *(rose, risen)* ansteigen; aufsteigen
roll [rəʊl] rollen
round trip Rundreise
route [ru:t] Route, Strecke
royal ['rɔɪəl] königlich
rubber Gummi
ruin ['ru:ɪn] ruinieren, verderben
rule regieren
run *(Straße)* verlaufen; **running water** fließendes Wasser

S

saloon [sə'lu:n] Salon *(Aufenthaltsraum auf Schiffen)*
salt [sɔ:lt] salzen
scale [skeɪl] Skala, Maß
scoop wheel ['sku:p wi:l] Schaufelrad
scorpion ['skɔ:piən] Skorpion
scout [skaʊt] Pfadfinder/in
seat [si:t] Sitz
second biggest zweitgrößte(r, s)
secret ['si:krət] geheim
section ['sekʃn] Abschnitt, Teil
sell [sel] *(sold, sold)* verkaufen
sender ['sendə] Absender/in
separate ['sepəreɪt] trennen
separation [ˌsepə'reɪʃn] Trennung
serious ['sɪəriəs] ernst, ernsthaft
serum ['sɪərəm] Serum
serve [sɜ:v] servieren
set *(set, set)* festsetzen; **set sail** auslaufen
settee [se'ti:] Sofa
settle ['setl] sich ansiedeln
settler Siedler/in
sew [səʊ] *(sewed, sewn)* nähen
sewage ['su:ɪdʒ] Abwasser
sewn [səʊn] *Part. Perf. „sew"*
shallow ['ʃæləʊ] *(Wasser)* flach
shark [ʃɑ:k] Hai
sheet [ʃi:t] Tuch
shelter ['ʃeltə] Schutzhütte
shield [ʃi:ld] Schild
ship verschiffen, versenden; **ship's biscuit** Schiffszwieback; **ship's carpenter** Schiffszimmermann
shoot [ʃu:t] *(shot, shot)* schießen
shovel ['ʃʌvl] Schaufel
sick [sɪk]: **I feel sick.** Mir ist schlecht.
side [saɪd] Seite
silage ['saɪlɪdʒ] Silofutter, Silage
silver ['sɪlvə] Silber
similar ['sɪmɪlə] ähnlich
since [sɪns] seit
sir [sə] *(Anrede)* mein Herr
site [saɪt] Stätte
situated ['sɪtʃueɪtɪd]: **be situated** gelegen sein
sketch ['sketʃ] Skizze
skiing ['ski:ɪŋ] Ski fahren
slave [sleɪv] Sklave, Sklavin
sleeping bag Schlafsack
sliding roof [ˌslaɪdɪŋ 'ru:f] Schiebedach
slimy ['slaɪmi] schleimig
slope [sləʊp] Hang
smell [smel] **1.** riechen; **2.** Geruch
smile (at sb.) [smaɪl] (jn. an)lächeln
smoke [sməʊk] Rauch
smooth [smu:ð] glatt
snakebite ['sneɪkbaɪt] Schlangenbiss
snow [snəʊ] Schnee
snowfall Schneefall

so: **… and so are you.** … und du bist es auch.
soap [səʊp] Seife
soft [sɒft] weich
sold [səʊld] *si. past/Part. Perf. „sell"*
sound Klang
speed [spi:d] Geschwindigkeit
spend [spend] *(spent, spent)*: **spend time/money** Zeit verbringen/Geld ausgeben
spent [spent] *si. past/Part. Perf. „spend"*
spice [spaɪs] Gewürz
spill [spɪl] verschütten
split (into sth.) [splɪt] (in etwas) aufteilen
sportswoman, *pl* **sportswomen** Sportlerin
spot [spɒt] Fleck
spread [spred] *(spread, spread)* ausbreiten, verteilen
sprinkle ['sprɪŋkl] streuen
square [skweə] **1.** quadratisch; **2.** Quadrat-
squash [skwɒʃ] zerquetschen
squeeze [skwi:z] zwängen
stage [steɪdʒ] Phase, Stadium
stairs [steəz] Treppe
star Stern
steady ['stedi] gleichmäßig, beständig
step Schritt; Stufe; **step back** zurücktreten
stick Stock
sticky tape [ˌstɪki 'teɪp] Klebestreifen
still still, ruhig
stir [stɜ:] (um)rühren
stocking ['stɒkɪŋ] Strumpf
store [stɔ:] lagern
straw [strɔ:] Stroh
strawberry ['strɔ:bəri] Erdbeere
strong [strɒŋ] stark
study ['stʌdi] studieren; gründlich ansehen
successful [sək'sesfl] erfolgreich
suck (up) [sʌk] (auf)saugen
sugar-loaf ['ʃʊgələʊf] zuckerhutförmig
sunscreen ['sʌnskri:n] Sonnenschutzmittel
sunshine ['sʌnʃaɪn] Sonnenschein
surprise [sə'praɪz] überraschen
surround [sə'raʊnd] umgeben
survival [sə'vaɪvl] Überleben
survive [sə'vaɪv] überleben
swallowtail ['swɒləʊteɪl] Schwalbenschwanz *(Schmetterlingsart)*
swear [sweə] *(swore, sworn)* fluchen
sweet [swi:t] **1.** süß; **2.** Süßigkeit, Bonbon
sword [sɔ:d] Schwert
swore [swɔ:] *si. past „swear"*
symbol ['sɪmbl] Symbol
symbolize ['sɪmbəlaɪz] symbolisieren
symptom ['sɪmptəm] Symptom

T

table ['teɪbl] Tabelle
tablespoon ['teɪblspu:n] Esslöffel
take *(took, taken)*: **take a look at sb./sth.** sich jn./etw. ansehen; **take part in** teilnehmen an; **take place** stattfinden
talk: **give a talk** einen Vortrag halten
task Aufgabe
teaspoon ['ti:spu:n] Teelöffel
tell sb. to do sth. jn. auffordern, etwas zu tun; jn. sagen, dass er/sie etwas tun soll
temperature ['temprətʃə] Temperatur
temple ['templ] Tempel
terror ['terə] Schrecken
theatre ['θɪətə] Theater
thick [θɪk] dick
thin [θɪn] dünn
thirst [θɜ:st] Durst
threat [θret] Bedrohung
throughout [θru:'aʊt]: **throughout the year** das (ganze) Jahr hindurch
thumb [θʌm] Daumen
thunderstorm ['θʌndəstɔ:m] Gewitter
tie [taɪ] (fest)binden
timeline Zeitleiste
times Zeitalter, Epoche
tired ['taɪəd] müde
to an
tonne [tʌn] Tonne
top: **at the top left/right** oben links/rechts; **top hat** Zylinder
topping ['tɒpɪŋ] Belag, Auflage
total ['təʊtl] Gesamt-
touch [tʌtʃ] berühren
track [træk] Weg, Pfad
tractor ['træktə] Traktor
trade [treɪd] Handel
trading wherry *trad. Handelsschiff der Norfolk Broads*
train [treɪn] trainieren, üben
transfer [træns'fɜ:] übertragen
translate [træns'leɪt] übersetzen
travel agency ['trævl eɪdʒənsi] Reisebüro
treasure ['treʒə] Schatz
trial ['traɪəl] Test
tricorn ['traɪkɔ:n] Dreispitz
tunic ['tju:nɪk] Kittel; Überwurf
turtle ['tɜ:tl] Wasserschildkröte
turn (sich) verwandeln
typical ['tɪpɪkl] typisch

U

unhappy [ʌn'hæpi] unglücklich
unicorn ['ju:nɪkɔ:n] Einhorn
union ['ju:niən] Union, Gemeinschaft
unit: **pig unit** Schweineställe
united [ju:'naɪtɪd] vereint
unsweetened [ˌʌn'swi:tnd] ungesüßt
unusual [ʌn'ju:ʒuəl] ungewöhnlich
up to bis (zu)
use [ju:s] Benutzung, Gebrauch, Verwendung
useful ['ju:sfl] nützlich

V

Valentine's Day ['væləntaɪnz deɪ] Valentinstag
valley ['væli] Tal
vanilla [və'nɪlə] Vanille
vapour ['veɪpə] Dampf, Dunst
vegetable oil ['vedʒtəbl ɔɪl] Pflanzenöl
vent [vent] Abzug
view [vju:] Aussicht
viewer ['vju:ə] Zuschauer/in
Viking ['vaɪkɪŋ] Wikinger/in
vinegar ['vɪnɪgə] Essig
violet ['vaɪələt] Veilchen
visibility [ˌvɪzə'bɪləti] Sicht
volcanic [vɔl'kænɪk] vulkanisch
volcano [vɔl'keɪnəʊ] Vulkan
voyage ['vɔɪɪdʒ] Reise, Fahrt
vulture ['vʌltʃə] Geier

W

war [wɔ:] Krieg
warship ['wɔ:ʃɪp] Kriegsschiff
washing-up liquid [wɔʃɪŋ'ʌp lɪkwɪd] Geschirrspülmittel
water ['wɔ:tə] gießen, bewässern
waterproof ['wɔ:təpru:f] wasserdicht
waterway Wasserstraße
wedding ['wedɪŋ] Hochzeit
weed [wi:d] Unkraut
weeding (Unkraut-)Jäten
weigh [weɪ] wiegen; **weigh down** niederdrücken
werewolf ['weəwʊlf], *pl* **werewolves** ['weəwʊlvz] Werwolf
wetland ['wetlənd] Feuchtgebiet
wheat [wi:t] Weizen
wherry ['werɪ]: **trading wherry** *trad. Handelsschiff der Norfolk Broads*
while [waɪl] während
whipped cream [wɪpd 'kri:m] Schlagsahne
wholesale ['həʊlseɪl] Großhandels-
wide [waɪd] breit; **wide-necked** breithalsig
wig [wɪg] Perücke
will [wɪl] werden
wine [waɪn] Wein
wise [waɪz] weise
wood [wʊd] Holz
woodland ['wʊdlənd] Wald(land)
wool Wolle
worker ['wɜ:kə] Arbeiter/in
would würde, würdest, würden
wrap [ræp] einwickeln

XYZ

yacht [jɒt] Jacht

List of names

Africa ['æfrɪkə] Afrika
Atlantic Ocean [ət,læntɪk 'əʊʃn] Atlantischer Ozean, Atlantik
Bligh [blaɪ]
Brazil [brə'zɪl] Brasilien
Caecilius [si:'sɪljəs]
Caius ['kaɪəs]
California [,kælə'fɔ:niə] Kalifornien
Celsius ['selsiəs]
Hallowe'en [,hæləʊ'i:n]
Herculaneum [,hɜ:kjʊ'leɪniəm]
India ['ɪndiə] Indien
Indian Ocean [,ɪndiən 'əʊʃn] Indischer Ozean
Jamaica [dʒə'meɪkə] Jamaika
Libya ['lɪbɪə] Libyen
Majorca [mə'jɔ:kə] Mallorca
Misenum [mɪ'seɪnəm]
Mount Kilimanjaro [maʊnt,kɪlɪmən'dʒɑ:rəʊ] Kilimandscharo
Mount Vesuvius [maʊnt,və'su:viəs] Vesuv
Norfolk Broads [,nɔ:fək'brɔ:dz]
North Shropshire [,nɔ:θ'ʃrɒpʃə]
Norwich ['nɒrɪdʒ]
Palaestra [pə'laɪstrə]
Plinius ['plɪnɪəs]
Pompeii [pɒm'peɪi] Pompeji
Pomponianus [,pɒmpəʊnɪ'eɪnəs]
Portsmouth ['pɔ:tsməθ]
Potter Heigham ['pɒtə'heɪəm]
Publius ['pʌbliəs]
Secundus [sə'kʌndəs]
Siberia [saɪ'bɪəriə] Siberien
Sicily ['sɪsəli] Sizilien
Stabiae ['steɪbɪaɪ]
Tanzania [,tænzə'ni:ə] Tansania
Tenerife [,tenə'ri:f] Teneriffa
Thames [temz] Themse
Theodosius [,θi:ə'dəʊsiəs]
Titanic [taɪ'tænɪk]
Warsaw ['wɔ:sɔ:] Warschau

Some numbers and measurements

1.6 one point six
207.53 two hundred and seven point five three
368.03 three hundred and sixty-eight point o three

1/4 a quarter
1/2 a half
1 1/2 one and a half
1 3/4 one and three quarters

1,041 one thousand and forty-one
33,912 thirty-three thousand nine hundred and twelve

30°C thirty degrees Celsius
-5°C minus five degrees Celsius

11.5% eleven point five per cent

Theodosius I Theodosius the First
Pope Paul II Pope Paul the Second
World War II World War Two *or* The Second World War

millimetres (mm)
centimetres (cm)
metres (m)
kilometres (km)

inches (in) *(1 inch = 2.54 cm)*
feet (ft) *(1 foot = 30.48 cm)*
miles (m) *(1 mile = 1.609 km)*
acres *(1 acre = 0.405 ha)*

hectares (ha)

square metres/kilometres (sq m/km, m^2/km^2)
miles per hour (mph)

grams (g)
kilos/kilograms (kg)
tonnes

millilitres (ml)

9 × 9 nine by nine

Bildquellen

Antippi Estate, Tanzania/Farming and Countryside Education, Warwickshire (S. 20: Fotos mit freundlicher Genehmigung v. Herr und Frau P. Chauhan, Antippi Estate und mit Unterstützung v.: www.face-online.org.uk); Archiv für Kunst und Geschichte, Berlin (S. 22–26: Marginalie; S. 41 unten; S. 43); Archivio Di Stato Di Lucca (S. 22 Illustration v. Giovanni Seracambi A.S.L., Biblioteca Manoscritti, n. 107, vignetta CXXXIII); Argus, Hamburg (S. 16 Steak: Schwarzbach; Karotten: Dott; S. 19 Nr. 1: Brigand); Broads Authority, Norwich (S. 36–40: Marginalie; S. 36–37 unten; S. 37; S. 38 oben; S. 38–39 unten; S. 40 oben links); British Tourist Authority, London (S. 41 oben); Bürgermeister von Warschau/ Ryszard Miklinski, Warschau (S. 7 oben); Corbis/Stockmarket, Düsseldorf (S. 14 Nr. 3: Ned Gillette); Corbis/PicturePress, Hamburg (S. 3–7: Marginalie; S. 3 Nr. 4: Morton Beebe; S. 6 Nr. 2: Morton Beebe; S. 6 Nr. 4: Eye Ubiquitous; S. 6 Nr. 5: Archivio Iconografico; S. 6 Nr. 6: Gregor Schmid; S. 7 unten links: Paul Almasy; S. 7 Mitte: Carmen Redondo; S. 7 unten rechts: Dave G. Houser; S. 14–15 Regenbogen: Galen Rowell; S. 17 Mitte: Nik Wheeler; S. 18 Milch: Eye Ubiquitous; S. 29 oben: David Turnley); Corel (S. 3 Nr. 1, 2, 3; S. 5 unten; S. 6: Nr. 1; S. 14 Nr. 1; S. 16–21: Marginalie; S. 16: Orangen, Äpfel, Kartoffeln, Spinat; S. 18: Bananen, Orangen; S. 21 oben; S. 30–35: Marginalie; S. 30 oben; S. 44–47: Marginalie; S. 45); Gareth Davies (S. 19 Nr. 2: Foto mit freundlicher Genehmigung v. Oliver Dowding und Farming and Countryside Education, Warwickshire); Dpa, Frankfurt (S. 6 Nr. 3: Bernd Heintze; S. 29 Mitte rechts: Curtis); Angelika Fischer, Berlin (S. 41–43: Marginalie); Gaff Fotoarchiv, Berlin (S. 17 Mitte rechts: Boris Geilert); Gettyimages/ Bavaria, München (S. 27–29 Marginalie: Vincent Besnault); Gettyimages/Stone, München (S. 14–15 Marginalie: Ralph H. Wetmore II; S. 14 Nr. 2: Robert Daly); Gettyimages/Photodisc, München (S. 14 Nr. 4: Steve Mason); Image Bank, Berlin (S. 3 Nr. 5: Terry Williams; S. 8–13 Marginalie: Joanna Mc Carthy); Imageoffice, München (S. 19 Nr. 3); Imagine, Hamburg (S. 34–35 unten: Ron Watts); Kingfisher Publications Plc, London (S. 12. Ilustration v. Nick Canaan in. *The Roman World/Kingfisher ancient world* by M. C. Corbishley); kpa, Köln (S. 29 Mitte links); Ladybird Books Limited, London (S. 44); Mauritius, Berlin (S. 3 Nr. 6: Haag + Kropp); The Madison Press Limited, Toronto (S. 8: Illustration v. Laurie McGaw © 1990 from *Secrets of Vesuvius*, A Tessloff/Madison Press Book); John Murray (Publishers) Ltd, London (S. 23–25: Illustration v. Anne Moore in: *Medieval Realms Special Needs Support Material*); Photovault/Images.de (S. 16: Pizza; S. 18: Mais); Polnisches Fremdenverkehrsamt, Berlin (S. 7 Mitte rechts); Reuters New Media Inc., München (S. 29 unten); Stockfood, München (S. 16 Käse, Spaghetti, Reis, Kuchen: S. & P. Eising; S. 16 Chips, S. 18 Tee: Susie Eising; S. 16 Burger: Gerhard Bumann; S. 16 Schokolade: Maximilian Stock LTD; S. 16 Joghurt: Molly Hunter; S. 16 Cornflakes: Thom DeSanto; Grillwürstchen: Jan-Peter Westermann); Visum/plus 49 GmbH, Hamburg (S. 17 oben: HHVW; S. 17 unten rechts: Gerd Ludwig; S. 18 Oliven, Käse-Sortiment: Guenter Beer); Peter Widmann, Tutzing (S. 16: Salat, Eis; S. 18: Kaffee); Carolin Wippel, Werner-Heisenberg-Gymnasium Bad Dürkheim, 6a (S. 34 oben); zefa visual media, Hamburg (S. 6 Nr. 7: K. Oster; S. 16: Müsli)

Titelbild

Pompeii (großes Bild): Roger Ressmeyer/ Corbis; Kaffeeplantage (oben links): variopress, Bonn; Zeichnung (unten Mitte): Carolin Wippel, Werner-Heisenberg-Gymnasium Bad Dürkheim, 6a; Pest (unten rechts): Archivio Di Stato Di Lucca (Illustration v. Giovanni Seracambi A.S.L., Biblioteca Manoscritti, n. 107, vignetta CXXXIII);

Textquelle

S. 23–25 adapted from "Alice's Wedding" in *Medieval Realms Special Needs Support Material* by Anne Moore. © John Murray (Publishers) Ltd, London